Praise for

The Master Builder

"An ingenious argument. A rich, detailed exploration
vitality of cells."

—*Kirkus*

"In *The Master Builder*, Alfonso Martinez Arias makes a
important, and compelling case for why an understan
living organisms must start with the cell. He offers a v
life that shows it to be much more interesting and in
than any simplistic notion of genetic blueprints can pro

—PHILIP BALL, author of *Critical Mass*

"The essence of science is that we never stop asking: Do
clearly, or have we fooled ourselves into certainty? In *Tl
ter Builder*, we follow that question into the cell, where
is said to rule. What Martinez Arias has found is muc
interesting: cells themselves, which we inherited from
line of ancestors stretching back to the earliest life, are
as integral to creating who we are. This book makes a n
stunning argument, not so much that we should put D
its place, but that we can see the grandeur of life as it t

—AZRA RAZA, author of *The First Cell*

"What came first, the chicken or the egg? *The Master*
poses a different question: What drives biology? Genes
His surprising answer shines new light on the fascinin
of development and offers a majestic cell's-eye view of lif

—LEE BILLINGS, author of *Five Billion Years of Solitu*

THE
MASTER
BUILDER

THE MASTER BUILDER

How the New Science of the Cell
Is Rewriting the Story of Life

ALFONSO MARTINEZ ARIAS

BASIC BOOKS
NEW YORK

Basic Books
Hachette Book Group
1290 Avenue of the Americas, New York, NY 10104
www.basicbooks.com

First Edition: August 2023

Published by Basic Books, an imprint of Hachette Book Group, Inc. The Basic Books name and logo is a trademark of the Hachette Book Group.

The Hachette Speakers Bureau provides a wide range of authors for speaking events. To find out more, go to www.hachettespeakersbureau.com or call (866) 376-6591.

Basic books may be purchased in bulk for business, educational, or promotional use. For more information, please contact your local bookseller or the Hachette Book Group Special Markets Department at special.markets@hbgusa.com.

The publisher is not responsible for websites (or their content) that are not owned by the publisher.

Illustrations by Selma A. Serra with the exception of Figures 4, 6, 8, 19, 24 (public domain), Figure 26 (modified from an image courtesy of Bernadette S. de Bakker), Figure 30 (courtesy of Prisca Liberali and Koen Oost), Figure 31 (courtesy of Matthias Lutolf), Figure 32 (courtesy of Jenny Nichols), Figure 33 (courtesy of Giorgia Quadratto), Figure 34 (courtesy of Madeline Lancaster), Figure 36 (courtesy of Jacob Hanna), Figure 37 (courtesy of Miguel Concha), and Figure 38 (courtesy of Nicolas Rivron).

Library of Congress Cataloging-in-Publication Data
Names: Martinez Arias, Alfonso, 1955– author. Title: The master builder : how the new science of the cell is rewriting the story of life / Alfonso Martinez Arias.
Description: First edition. | New York, NY : Basic Books/Hachette Book Group, 2023. | Includes bibliographical references and index.
Identifiers: LCCN 2022049294 | ISBN 9781541603271 (hardcover) | ISBN 9781541603288 (ebook)
Subjects: LCSH: Cells. | Life (Biology)
Classification: LCC QH581.2 .M383 2023 | DDC 571.6—dc23/eng/20230418
LC record available at https://lccn.loc.gov/2022049294

ISBNs: 9781541603271 (hardcover), 9781541603288 (ebook)

To my parents, for making me believe

CONTENTS

INTRODUCTION

For you formed my inward parts;
you knitted me together in my mother's womb
I praise you,
for I am fearfully and wonderfully made.

—PSALM 139: 14–15

E VERY ANIMAL AND PLANT ON EARTH HAS AN AWESOME beauty: the majesty of an oak, the delicate fabric of a butterfly, the grace of a gazelle, the imperious presence of a whale, and, of course, us—humans—with our mixture of wonders and fatal flaws. Where does it all come from? History is full of stories in reply to this question. For example, in Mayan tradition the answer is corn; other cultures suggest various forms of egg as the source. In many, the origin is some claylike material shaped by the might and imagination of a powerful entity that breathes life into it. From such starts, multiplication follows, and the earth is populated, though the details of how this happens are scant.

Over the past century we have discovered a material explanation for the source of life, one that needs no divine intervention and provides a thread across eons of time for all beings that exist or have ever existed: deoxyribonucleic acid (DNA). In the words of the US National Human Genome Research Institute, "Your genome [DNA] is the operating manual containing all the instructions that helped you develop from a single cell into the person you are today." And yet, while there is little doubt that genes have something to do with what we are and how we come to be, it is difficult to answer precisely the question of what their exact role in all of this is.

A closer look at how genes work and what they can accomplish, compared to what they are said to achieve, casts doubt on the assertion that the genome in particular contains an "operating manual" for us or any other living creature. When it comes to the creation of organisms, we've overlooked—or, more accurately, forgotten—another force. This book is about the origin and power of that force: our cells.

WHAT MAKES YOU AND ME INDIVIDUAL HUMAN BEINGS IS NOT A unique set of DNA but instead a unique organization of cells and their activities. The story of Karen Keegan, a fifty-two-year-old woman in desperate need of a new kidney, is an example.

After consulting with doctors, Karen knew that a donor's kidney would have to be as close a genetic match as possible to reduce the chances that her immune system would reject it as a foreign invader. She was lucky, the doctors had told her. As the mother of three adult sons, she was very likely to find a match within her immediate family. By the rules of genetic inheritance, each of her kids would share about half of his DNA with her, which would make them all good donors. It was just a matter of doing a blood test to see which son was her best match based on exactly which DNA he'd inherited from her. But when the test results arrived from the lab, Karen was in for a shock: two of her three sons could not be hers, the doctors said, because they did not share enough DNA. There had to have been a mistake in the test, Karen protested. She'd been pregnant and given birth to all three of her sons; she had felt them growing (and kicking!) inside her.

Lynn Uhl, a specialist at the hospital, knew Karen, and she knew that Karen had given birth to the children. The chances that not just one but two of Karen's sons had been mistakenly switched at birth was also astronomically unlikely. So too was the chance that there had been a mix-up in the blood lab. On a hunch, Uhl decided to check Karen's blood sample against some tissue from another part of Karen's body. This test solved the riddle: Karen did not have one DNA sequence, or genome, in her cells. She had two.

Fifty-three years before, early in Karen's mother's pregnancy, two separate eggs had been independently fertilized, giving rise to two separate balls of cells, each with its own DNA. At some point in the rush of cell division and multiplication that follows fertilization of the egg by sperm, the two groups of cells fused into one. Instead of developing into twins, they developed into Karen, with cells from both balls randomly distributed throughout her one body. While most of Karen's body had cells from one of the groups, it just so happened that two of her sons came from eggs that had been generated by the other.

People who carry more than one complete genome are called *chimeras*, after the fire-breathing lion of Greek mythology with the head of a goat growing out of its back and the head of a snake growing out of its tail. The term denotes that they are combinations of more than one creature. Karen is not alone in being a natural chimera. Indeed, the first human chimera was identified in 1953, the same year that the double helix structure of DNA was discovered. And today, some scientists estimate that about 15 percent of people are chimeras. Sometimes only blood cells are mixed up, but other times, as in Karen's case, two separately fertilized eggs start to develop and then get fused together.

Ever since the day in 1953 when James Watson and Francis Crick unveiled their model of the double helix to account for the structure of DNA, we've been in thrall to genes. We think of every aspect of ourselves as being determined by our DNA, from the color of our eyes to our propensity for a particular disease. In the minds of some, DNA even sets the parameters for a person's intellectual ability or temperament: *It's in her genes*, a parent will say about a child. We take a swab of cells from the cheek and get our DNA tested to learn "who we are," as if tracing which genes we inherited from whom tells us anything about ourselves right now. DNA has become so central to our sense of identity that we even use it as a metaphor for social organizations: *It's in our DNA as a company*, a CEO will say, or *as a team*, says a coach. Yet chimeras are just one way in which nature shows us that DNA does not define who we are. Karen is not defined by a DNA sequence; she has two.

The publication of the human genome ushered in an era in which people think that most noninfectious diseases have some genetic basis, underscoring the connection between DNA and us. For conditions linked to a mistake in a single gene—like cystic fibrosis, hemophilia, or sickle cell anemia—a focus on DNA will almost certainly allow scientists to develop cures. Along these lines recently, cutting-edge technologies like CRISPR—the so-called genetic scissors that allow editing of DNA at will—have thrown up a host of potential treatments. For example, gene-editing interventions using CRISPR (an acronym for "clustered regularly interspaced short palindromic repeats") have been shown to repair a single change in the DNA for the beta-globin gene (*HBB*) that produces sickle cell anemia and thereby restore the health of individuals. Other cases are in the pipeline.

But even in instances like this, there are problems. The relationship between changes in a gene and a dysfunction is not usually as straightforward as in the case of sickle cell anemia. Having mutations in the breast cancer type 1 (*BRCA1*) or type 2 (*BRCA2*) gene makes it more likely that the body can't produce the functional proteins needed to effectively destroy cancer cells in breast tissue, but it does not say you will get cancer. Mapping gene mutations to cell malfunctions may help us to understand what happens when a gene is faulty or absent, but more often than you think, the observation doesn't tell us how cells use the normal form of the gene to make normal tissues and organs. In fact, over 60 percent of birth abnormalities cannot be linked to specific genes. Many chronic diseases are caused not by genetic predisposition but by how cells respond to their environment—including in breast cancer, where only 3 percent of people diagnosed have a mutation in their *BRCA1* or *BRCA2* gene.

Of course, genes do carry information that contributes to our being. Identical twins are the classic example since they share all of their DNA at birth and look uncannily similar. At the same time, identical twins raised in the same home can develop different personalities, different medical conditions, and, sometimes, different physical traits. The question is not whether DNA has something to do with the way we look or behave but rather what exactly its role is.

It's strange how completely we've given in to a gene-centric view of life. We've been aware of the workings of cells for well over a century, and through years of study, we have come to know their content and organization in detail. Some we know as essential functional entities. The immune system comprises an army of cells that fight infections and heal injuries, while neurons process information to generate and control our movements and thoughts. Recent advances in our ability to scrutinize cells' contents and activities have revealed them to be dynamic entities capable of creating and destroying time and space. We have filmed their interactions and observed how they work in groups to build and maintain organisms. We have learned that our bodies are in constant flux because the cells that make them up are themselves in constant flux. When we consider life from the perspective of the cell, the result is a breathtaking vista of spatial and temporal choreographies.

I have devoted my career to studying how cells come together to generate organs and tissues in animals from fruit flies to mice to human beings. I trained as a geneticist, and for much of my career I was a professor in the Department of Genetics at the University of Cambridge using the science of genes to search for answers to biology's big questions. But I have grown increasingly uneasy about how much genes are blamed for things that they have nothing to do with. Genetics has provided important glimpses into the processes of animal and plant development, but we have overstretched what genes can explain.

The reason is simple. Geneticists have been so successful at finding changes in genes associated with dysfunction that we've fallen into the trap of equating correlation with causation. We've transformed method into explanation. We have turned tools for studying life into the architects and builders of life. However, if we remove a few bricks from a key position in a house and the house falls down, we don't suddenly think the bricks are the house's blueprint or its architects. So why do we think that if we remove a gene from the genome and the organism stops developing or functioning, genes are the blueprint or architects of life? As famed French mathematician Henri Poincaré might have phrased it, cells are no more piles of genes than a house is a pile of bricks.

Many will say that there is nothing here to challenge the gene-centric view of development and evolution. After all, cells are an inevitable consequence of the activity and interactions of the genes that lie in their genomes. There is some truth in this, but the fact is that cells have powers that DNA cannot dream of. DNA cannot send orders to cells to move right or left within your body or to place the heart and the liver on opposite sides of your thorax; nor can it measure the length of your arms or instruct the placement of your eyes symmetrically across the midline of your face. We know this because each and every cell of an organism generally has the same DNA in it, with the same monotonous structure. As we shall see, cells can in fact send orders, measure lengths, and much more beyond that. In chimeras such as Karen Keegan, cells negotiate the differences between the two genomes coming together to create one body. To do their masterful handiwork, cells *use* genes, choosing which will or will not be turned on and *expressed* to determine when and where the products of genes are deployed. An organism is the work of cells. Genes merely provide materials for their work.

The view of biology I share in this book has grown with me over the years but became obvious through experiments done in my lab, as well as a few others, in which cells have displayed astonishing abilities. Our experiments started by trying to understand why cells behave differently in culture versus in embryo. We found that when a particular type of mouse *embryonic stem cells*—that is, cells that can give rise to any type of organ or tissue—are left to roam on a Petri dish in certain conditions, they will become different from each other; they generate the different types of cells that make up the embryo but do so in a disorganized manner. If the same cells, with the same genes, are placed in an early embryo, however, they will faithfully contribute to the embryo. Same cells, same genes. So, something other than genes must be involved in making an embryo. We went on to prove this by developing conditions in the lab in which the cells will imitate many of the processes that lead to the first organization of a body plan in an embryo. The ability to use cells to build structures resembling tissues and organs and even embryos in the lab represents the birth of a new kind of engineering, one that allows cells to show us what they need to build organisms, using their tools and following their rules.

Through this research, I have come to recognize a creative tension between genes and cells that lies at the heart of biology. Cells don't merely multiply, regulate, communicate, move, and explore; they also count, sense force and geometry, create form, and even learn. You have never been just a gene or even a set of genes. Instead, you can safely trace your origins back to a first, single cell within your mother's womb. Once this first cell came into existence, it began to do things that are not written in DNA. As it multiplied, it created a space in which the emerging cells assumed identities and roles, exchanged information, and used their positions relative to each other to build tissues, sculpt organs, and eventually produce a whole organism—you.

THE PAGES THAT FOLLOW WILL INTRODUCE YOU TO CELLS—THEIR origins, their relationship with genes and with each other, and how they came to weave that crucible of the individual that we call the embryo. My narrative has three acts. In the first, after reviewing what genes are and how we have come to accept them as harbingers of fate, I introduce you to the cell and begin to explore its relationship to the gene. We shall see how, at a certain point in life's history, cells invented plants and animals by gaining the capacity to use genes to cooperate and communicate with each other on a permanent basis. I challenge the well-known view of biology from the perspective of the "selfish gene" and, as an alternative, present a cell's-eye view of our world. In the second part, we dive into the details of the relationship between cells and genes and learn about the language and techniques cells use to create embryos—sometimes in secret, as in our case. In these chapters, we shall see cells at work in the creation of embryos and learn about our individual origins, hidden in the wombs of our mothers. Finally, I tell you about a remarkable recent finding that you have not one single genome but many—as many as you have cells and probably more—destroying the notion that there is a strong and deep relationship between us and one genome. In the third part we learn how, from the perspective of the cell, we are a different being every year. I tell you about stem cells and share recent advances in the use of these magic instruments of bodily renewal and how their study is revealing a surprising potential that we can harness to reconstruct organs, tissues, and embryos in the lab. The cell's-eye view of life raises questions about human identity and nature that we need to address as a future beckons in which the manipulation of cells can create not only structures to repair our bodies but also, surprisingly, whole organisms and, perhaps at some point, beings like us.

The story this book has to tell is far from comprehensive. My intention is not to provide a crash course in the biology of cells or a scholarly account of how cells build organisms; rather my aim is to bring the cell to the foreground of ongoing debates about identity, health, and disease and to highlight the crucial role it plays in these aspects of our lives. For this reason, to achieve my goals, I have had to be concise in explanation and selective in examples; interested readers will find suggestions for further reading at the end of the book.

For one feature of my story, I do need to apologize: the extensive focus on animals at the expense of plants. This is due in part to my expertise as a developmental biologist who studies animal development and also to my interest in exploring the origin and identity of humans, which you, the reader, as a fellow human, may very well share. In addition, an emphasis on animal cells seemed particularly vital in the context of the global coronavirus pandemic. As I write these lines in Barcelona, SARS-CoV-2 appears to have loosened its grip on the world, and the reason for hope lies not in the biology of the virions, which are stretches of ribonucleic acid (RNA) in a protein coat, but in the biology of our own cells. The millions who have died of Covid-19 have been killed by cells overreacting to infection, and it is cells that save us from infection. The vaccines work by exposing immune cells to just enough of the virus to form a "memory" of it so that the next time anything with the same stretch of RNA enters your body, your immune cells will destroy it and thwart its intent to trigger an attack of some cells on their own host. Because the technology to create the vaccines was new, all the headlines are about the power of ribonucleic acid, but we should not forget that it is our cells we should be thanking.

With many other diseases, the future holds great promise for developing treatments based on a fuller knowledge of how cells use genes. This will require a shift in how all of us—scientists and nonscientists—talk and think about life. DNA, genes, and CRISPR have rightly become part of our common language. Over the next few years, cells, embryos, and development should also become an important part of our vocabulary, because they pertain to where we come from, what we are, and what we become.

I set out to write this book with the idea of sharing a vision of the natural world gathered over my years studying the development of animals. As the project progressed, I realized that embryonic development challenges the dominant view that what we are starts and ends in the genes. From what I have seen in my own lab, as well as in decades of experiments with both genes and cells, it's become clear to me that the cell is not just a building block of our tissues and organs; it is also architect and builder. We cannot know or heal ourselves until we integrate this reality into our understanding of life.

PART I
THE CELL AND THE GENE

That a single cell can carry the total heritage of the complex adult, that it can in the course of a few days or weeks give rise to a mollusc or a man, is one of the great marvels of nature. In an attempt to attack the problems here involved we must from the outset hold fast to the fact that the specific formative energy of the germ is not impressed upon it from without, but is somehow determined by an internal organization, inherent in the egg and handed on intact from one generation to another by cell division. Precisely what this organization is, we do not know.

—E. B. WILSON, *THE CELL IN DEVELOPMENT AND HEREDITY*

I find it strangely liberating to think of genomes and emergent cell behaviours as two rather independent entities that need to remain compatible with each other in order to create a functional living system.

—PAVEL TOMANCAK, TO THE AUTHOR ON TWITTER

– one –

NOT IN THE GENES

I N MANY AIRPORTS AROUND THE WORLD, AS YOU GO THROUGH passport control, you're asked to place an index finger on a small box with a glass plate and a flashing light—a scanner that "reads" your fingerprint. This device is used because it's a more reliable way of confirming your identity than comparing your face to a passport photo. If time has passed since the photo was taken, you may have changed your hairstyle, gained or lost a few pounds, or developed smile lines around your mouth or worry lines across your forehead. At the very least you will have aged; alas, we all do. It's also possible to change one's facial appearance to look enough like someone else's photo to fool both humans and today's computers. But the patterns of ridges on your fingertips do not change. They'll be exactly the same as they were on the day you were born—actually, since *before* you were born. And no two people have the same fingerprints.

So, you might think that, because fingerprints are unique, we ought to be able to trace them to a gene, maybe two or three. After all, we've been told again and again that we're defined by our DNA, that genes are us. But the relationship between genes and fingerprints is not straightforward; even identical twins, who share 100 percent of their DNA at birth, don't have the same fingerprint patterns. In fact, every one of your ten fingers has its own fingerprint. That's why you can only unlock your mobile phone with one particular finger and the same finger of your opposite hand will not do.

Though genetic studies have found hundreds of genes that appear to be associated with fingerprints, with each gene making a small contribution to the patterns, even added together, their influence is not significant in defining the final fingerprint.[1] Genes contribute to those patterns not as determinants but as bit-part players in a larger process that designs and carves the furrows at the tips of your fingers; fingerprints are not written in the genes.

We think about features like our eye and hair color, the shape and size of our nose, or the length of our fingers as being linked to genes we've inherited from our parents, but in reality the contribution of genetics to these features is not the one we have been led to think. You inherit the basic color of your eyes from your parents— brown, blue, gray, or green—but if you look closely at your irises, you'll notice that they have an intricate pattern of rings, crypts, and furrows. Despite sharing the same genes, each eye has an entirely distinct and unique pattern, and, as with fingerprints, genes alone cannot explain how they came to assume their final form.

It is surprising that the biomarkers that we most often rely on to distinguish us from others are not written in our DNA. This is because your fingertips and the irises in your eyes are not made of or by genes; they are made of and by cells. If you look through a magnifying glass at your fingertips, you will see a delicate pattern of ridges sculpted over a fuzzy mat, underneath which, and not visible to your eyes, thousands of cells are stuck to each other. Those patterns were created when you were in the womb by the cells themselves using tools and materials provided by the genes.

How then have we come to believe that our being and our identity are found in our genes? To answer this question, we first need to understand how genes became the protagonists in that central narrative of how we come into being. We'll start at the beginning, taking a fresh look at what genes are, how they work, and why they have become a shorthand for much of what we are and how we come to be. It is a story that involves nucleic acids, proteins, and mutations, as well as creative scientists and their visions. Grasping these nuts and bolts of our existence is crucial to understanding the handiwork of cells.

RULES OF INHERITANCE

Over the course of history, people came to realize that significant aspects of our being are inherited. Dogs beget dogs, sheep beget sheep, and the ancient practices of animal breeding and agriculture hinge on an understanding that traits can be passed along to the next generation. Closer to home, some of you look like one of your parents or grandparents. Resemblances like this led to the suspicion that something central to the being of an organism is passed from generation to generation, and for a long time the vehicle was suspected to be blood. This notion was closely attached to royal dynasties, with their accolades and privileges passing with the blood from one generation to the next.

The notion of the inheritance of characters became stronger with the realization that not only appearance but other defined traits run in families. In 1751, the French polymath Pierre-Louis Moreau de Maupertuis published a pedigree with three generations of a family, many of whose members exhibited six fingers, leading to the conclusion that this trait was hereditary. This is probably the first formal account of the inheritance of a character. Similar observations of inheritance applied to some diseases, as in the case of hemophilia, first described in 1803 by the Philadelphia physician Joseph Conrad Otto as "familial hemorrhagic bleeding in males," which he traced to a woman who had settled in Plymouth, New Hampshire, in 1720.

As humans learned about heredity, we used its lessons in practical ways. For thousands of years, we have bred animals based on how they look, with the goal of improving the quality of their meat and hides to suit our needs and purposes. By the eighteenth century, farmers and graziers had learned enough about inheritance to conduct breeding on a large scale. The most influential of the breeders, a man called Robert Bakewell living at Dishley Grange in Leicestershire, collected sheep from around England to develop a variety that had splendid wool as well as a much meatier body. The cattle bulls that he bred were also bigger and meatier: in 1700, before the British agricultural revolution, the average bull at slaughter weighed 170 kilograms; a century later it weighed about 370 kilograms. Bakewell operated at the unit of the full animal, finding those specimens that had all the qualities he wanted and breeding them with each other. Any offspring that didn't match his desired type were simply removed from his breeding program. As successful as he was, this breeding method was based entirely on animals' observed traits—what scientists now call a *phenotype*.

That something was afoot with inheritance was common knowledge between plant and animal breeders, but uncovering what lay behind those pedigrees took the application of method, attention to detail, and counting peas.

The first scientific approach to understanding heredity was put forward in 1866, when the Moravian monk Gregor Mendel gave a set of lectures on the results of his experiments cultivating pea plants. Mendel had observed that when both parent plants had wrinkled peas, the offspring plants would have wrinkled peas too. If only one parent had wrinkled peas and the other came from a stable line of smooth peas, all the offspring would have normal smooth peas. Surprisingly, however, when he crossed these offspring with others of the same type, the wrinkles reappeared in a portion of the next generation. Mendel found this to be true for other traits too—whether the plant was tall or short, whether its pea pods were tight or balloon-like, or whether the flowers were purple or white. There were identifiable patterns in how traits were conveyed from parents to offspring, disappearing and reappearing through multiple generations. Moreover, these patterns of inheritance obeyed some numerical rules.

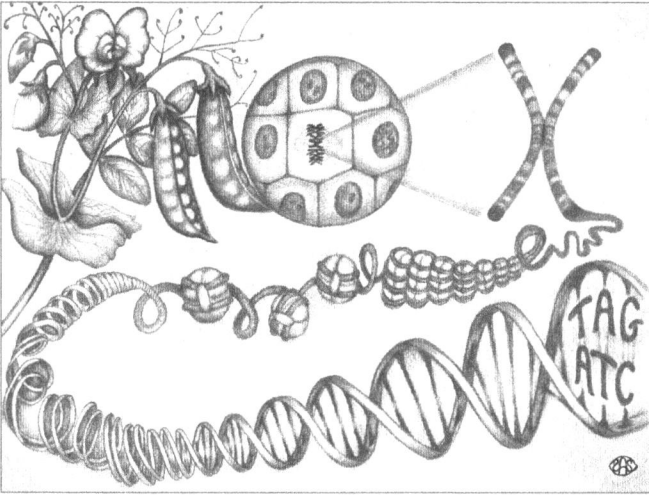

FIGURE 1. The experiments of Gregor Mendel with peas led to discovery of the genes, the chromosomes, and later their material nature: the DNA double helix.

This led Mendel to propose that a trait associated with a physical particle of some sort could be passed on from generation to generation. Plants, he said, had two such particles associated with every characteristic; one was inherited from the male parent and the other from the female parent. Some particles were *dominant*, and others were *recessive*, so that when a dominant particle and a recessive particle were paired, only the traits associated with the dominant particle were observed in the offspring. For example, smooth is dominant over wrinkled, so a pea can only be wrinkled if it has inherited wrinkled particles from both parents. This rule explained what he had observed but did not help him figure out the nature of the particles or identify the biological mechanism (other than sexual reproduction) at play in the patterns of inheritance he observed.

Mendel published his results with excitement, but only a handful of scientists paid notice. Indeed, it is said that the pages of Mendel's paper remained uncut and unread in the library of Charles Darwin. Not until the dawn of the twentieth century did researchers become aware of Mendel's work after observing similar patterns of

inheritance in a variety of plants. In particular, the British biologist William Bateson was enthralled with Mendel's experiments
and, together with Edith Saunders, engaged a group of women at
Newnham College, University of Cambridge, to carry out groundbreaking research into the details of the inheritance of traits. Study
after study with both animals and plants confirmed and elaborated
Mendel's rules of inheritance: individual traits were conveyed from
parents to offspring in the form of particles that, though mysterious
and invisible, followed precise and reproducible numerical patterns
of dominant or recessive.

In 1905 Bateson coined the word *genetics*—from the Greek word
genos, meaning *birth*—to describe the study of the inheritance of
traits. The term *gene* was chosen a few years later to refer to the
smallest unit of inheritance associated with a given trait that could
be observed and followed through generations—akin to the particle
that Mendel believed made peas wrinkled or smooth. The combination of genes associated with a specific trait came to be known as
the *genotype*, an invisible analog to the phenotype used to refer to
observed characteristics. Over the years, genetics morphed from the
study of inheritance into the study of the transmission and effect
of genes, even before anyone had seen a gene or knew what genes
consisted of. Once Mendel's rules of inheritance were firmly established, scientists turned to looking for these elusive particles. The
most natural place to search was in cells, the smallest unit of life.

When researchers looked at cells in detail under a microscope,
they saw an inner core, or nucleus. This looked something like
a vault, full of small, threadlike structures. Scientists called the
threadlike structures *chromosomes*—from the Greek *chromo*, meaning *color*, and *soma*, meaning *body*—because they could be stained
with certain dyes. By counting the number of chromosomes and
watching how this changed when plants and animals were crossed
to create offspring, it became clear that they were the most likely
home for what Bateson had named genes. Curiously, every organism
seemed to have a consistent number of chromosomes, which come
in pairs, as expected from Mendel's findings. For example, humans

have 23 pairs of chromosomes, or 46 in total; fruit flies have 4 pairs, or 8 in total; and the hermit crab has 127 pairs, with a grand total of 254. But the world record goes to certain ferns that have around 1,200 chromosomes. Furthermore, males and females differ in one of the pairs: in humans, females have two X chromosomes, whereas males have one X chromosome and a smaller Y chromosome. Could it be that chromosomes determined our differences—that Mendel's material particles reside there?

Now scientists knew where to look for genes, but there was still much to uncover. They suspected that phenotypes were associated with those structures in the nucleus that were passed on from parents to offspring in species-specific numbers, but how it all worked remained a mystery. If they could only sort out the complex chemical composition—proteins, acids, bases, even phosphorus, an unusual element in a living system—they would be able to decode the genetic blueprints locked up inside the nucleus of the cell.

WHAT'S "IN" A GENE?

In 1943, at the height of World War II, a bacteriologist working at the Rockefeller Institute in New York City was putting together a summary of his research over the previous twenty years. Its subject: the chemical nature of what he called the "transforming principle," a mysterious substance that appeared to convert pneumococcus, the bacterium that causes pneumonia, from a benign organism into a killer.

Oswald Avery had first become interested in pneumococcus in the 1920s, during efforts to develop a vaccine to avert another pandemic like the Spanish flu, which had devastated Europe and the United States at the end of the Great War. At the time, the influenza virus had not yet been identified, and pneumonia was considered to be public health enemy number one. Avery had become intrigued by the work of Frederick Griffiths in the United Kingdom's Ministry of Public Health, who found that two pneumococcus strains, dubbed R and S, behaved very differently when injected into a mouse. Strain R rarely caused pneumonia, while strain

S killed every exposed animal. However, if he mixed dead strain S cells with living strain R ones, every animal injected with the blend died too. Avery wondered what part of a dead cell could be so powerful as to transform the very character of living cells that were put in the same test tube.

Science is often a process of incremental discovery. Change one thing and see what happens. Start again; change another. Start over, again and again, until you observe a cause-and-effect relationship. One by one Avery removed components of the dead S cells to see which one turned the benign R cells into killers. A substance called *deoxyribose nucleic acid* (DNA) caught his attention. When he removed DNA from the debris of the S strain, the R strain stayed true to its old self; it didn't turn into a killer. Providing DNA to the R strain turned it into a killer. And with that, he had identified the stuff of genetics along with its ability to change the properties of a living organism.

DNA determined the fundamental characteristics of pneumococcus. In a letter to his brother, Avery considered the implications of these findings:

"What is the chemical nature of the transforming principle? . . . Of course, the problem bristles with implications. . . . It touches genetics, enzyme chemistry, cell metabolism and carbohydrate synthesis, etc. Today it takes a lot of well documented evidence to convince anyone that . . . deoxyribose nucleic acid, protein-free, could possibly be endowed with such biologically active and specific properties." This was a significant leap in our understanding of inheritance. DNA appeared to be what comprised Mendel's elusive particles. More interestingly, these traits were not only inherited but could be transferred from cell to cell, transforming a cell's characteristics, outside breeding and reproduction.

Perhaps because, as in the case of Mendel, he had made such a big leap, few people paid attention to Avery's finding. One who did was James D. Watson, then a student at the University of Chicago. He had a hunch that Avery's transforming principle might harbor the secret to how living systems inherit distinct features.

In the early 1950s in the United Kingdom, Watson teamed up
with the physicist Francis Crick in Cambridge to uncover the physical
structure of DNA. At this time the basic chemistry of DNA was well-
known, but it looked too simple to account for the variety of life on
earth. DNA was made of the sugar ribose, phosphate, and four chemical compounds—adenine (A), cytosine (C), guanine (G), and thymine (T)—which are called *bases* because of their chemical nature.
The four bases appeared in specific proportions in different organisms.
Could these four letters be the secret of life on earth? The question
remained: How were these chemicals organized into a structure?

The answer came from a series of crisp X-ray photographs of
DNA fibers. These were taken by a skilled young scientist, Rosalind
Franklin, and examined—without Franklin's permission—by Watson and Crick from 1951 to 1953. Using models that would represent what was seen in the photographs, they teased out the iconic
double helix structure—two strands, within a basic backbone of
sugar and phosphate, twisting around each other, each a linear combination of As, Cs, Gs, and Ts. The strands mirrored each other,
and the bases were bonded in precise pairs: A bonded with T, and G
bonded with C. Where one strand had a stretch AGCT, its partner
in the helix had the stretch TCGA.

Watson and Crick are rightly associated with discovering the
structure of DNA (using Rosalind Franklin's data), but they did
something bolder and more consequential. In the double helix they
saw the answer to many of their predecessors' questions at once:
Mendel's particles, Bateson's genes, Avery's transforming principle,
and the basis for the mutations that geneticists worked with. The
two strands they described could come together—one the original and the other a mirrored copy—to reproduce cells and organisms, as the inheritance of one strand would serve as a template
for the other. The double helix provided an explanation for why
sparrows breed sparrows and not swallows, why blue whales breed
blue whales and not dolphins, and why sons resemble fathers and
mothers, grandfathers and grandmothers. The complementarity of
the two strands meant that every time a cell divided, every time an

organism reproduced, one of the strands could re-create the missing strand from a template (barring any errors, of course). They suggested that whatever defined a sparrow, swallow, whale, dolphin, you, or me was captured in this sequence of chemicals. The double helix laid the foundations for the gene-centric view of nature that has dominated our view of life for nearly a century.

And yet, this gene-centric view does not quite explain everything. Since the discovery of the structure of DNA, it's become common to refer to DNA as the "book of life," a text made up of a sequence of letters—As, Gs, Cs, and Ts—that serves as an instruction manual for building organisms. But what are the instructions for, and who carries them out?

DNA has much less in common with an instruction manual than we might think. When we picture such an instruction manual—for example, to make a piece of furniture—we might imagine a series of images showing you the pieces you need at each stage, with arrows pointing out how to assemble them into the bookshelf or cupboard you want. But where instruction manuals show us the order in which to do things, where to do them, and what to use to accomplish each step, DNA does nothing of the sort. Although the base pair letters are arranged along a string, there is no set order in which letters must appear. Certain stretches of letters along the string do, however, carry messages: these are what we call *genes*. Finding them is not straightforward.

For all we have learned about genetics, breaking strands of DNA down into individual units, genes, is not a simple matter. Talk to biologists, and they may very well tell you that it is nearly impossible to define what an individual gene is. To some, a gene is just a simple chemical structure, a sequence of DNA drawn out of combinations of those essential four letters that is inherited. But for many people, both scientists and nonscientists, a gene is specifically a unit of inheritance associated directly with a particular characteristic. The source of this disagreement over the definition of a *gene* arises from the awkward fact that only about 1 to 3 percent of your *genome*—that is, the total amount of DNA in each of your cells—is directly associated with inherited traits. This percentage varies from organism to organism, but across all species, only a proportion of the DNA is associated with genes. When it comes to the rest of the sequences of As, Gs, Cs, and Ts, we still don't know much about what it does. For this reason it's sometimes said that genes are like words tucked within long stretches of gibberish—though of course the genes are recognizable not as words but instead as chemicals.

Still, there is some truth to likening a gene to a "word"—like a word, genes do have meaning, and the longer the stretch of letters comprising a gene, the longer the message it conveys. Each position along the string can be filled by any of the four chemical bases—A, G, C, or T. For example, if a gene is four letters long, the gene could come in one of 4^4, or 256, different possible combinations, meaning it could have 256 different chemical meanings. If it is five letters long, it could come in 4^5, or 1,024, different combinations. If it's thousands of letters long, as is the general case, the possible combinations are practically infinite. However, as in any language, not all lengths and combinations of letters and words make sense.

To begin to comprehend what genes actually are and whether their role is really as important as the gene-centric view might suggest, scientists needed to dig deeper, decoding how the words and texts of a genome are defined, written, and read.

THE LANGUAGE OF GENES

Steve Jones, a geneticist at University College London and author of several popular books about genetics, starts his courses on genetics by telling students that his job is to make sex boring. The discipline of genetics, he says, is sex without the fun. He is probably right. Sex is the essence of genetics, but though there is a kind of sex involved in crossing pea plants and mice, there is nothing very provocative in counting classes of peas, accounting for color coats, or calculating exponentials, and certainly not in muttering sentences like AAAGTCCCTTA.

Ultimately, the genome is less a steamy romance novel than a puzzling literary text in need of decoding. In addition to referring to DNA as the "book of life," or a script, scientists commonly talk about messages in DNA being *transcribed* into an intermediary, a messenger between the gene and the trait. Another nucleic acid, *ribonucleic acid* (RNA), serves this role of intermediary. The molecule of RNA is very similar to DNA, with two differences: in place of the sugar deoxyribose, it has ribose, and in place of the base thymine (T), it has uracil (U). Because of these molecular differences, RNA does not form a double helix like DNA; instead, it forms either linear ribbons or complex, three-dimensional structures.

During transcription, stretches of the DNA double helix are "unzipped" from a chromosome and laid out as two separate strands. The parts of the DNA that are needed for a particular task are copied into a ribbon of RNA, which, unlike DNA, can leave the chromosome. This is how genes are *expressed*, shorthand for transcribed. Whereas there is only one copy of the gene in the DNA, many copies are made of the ribbon of RNA. In addition, whereas DNA exists for the entire life of the cell and can survive even after the organism dies—as Oswald Avery discovered with his killer S pneumococcus—RNA has a short life. Once the RNA copies are made, the DNA is zipped back up again to minimize the chances of losing the information contained within it.

Some RNA molecules play important roles in controlling what DNA does, including when and where genes are expressed. More significantly, it is through RNA molecules that the meaning of the "words" encrypted in DNA is decoded. This is done by taking the message in RNA and turning it into a protein. These types of RNA molecules are called messenger RNA (mRNA).

Proteins are the workhorses of the cell and come in many varieties. One class of proteins—enzymes—performs chemical reactions that mediate processes like digestion, where food is broken down into molecular particles to generate energy, and immune responses, where toxins are broken down into inoffensive waste. Other proteins have more structural roles: the protein Keratin provides structural support and protection to cells and makes up the bulk of your hair and nails, while hemoglobin inside your red blood cells moves oxygen around the body. The protein dystrophin serves as a flexible binding agent, connecting cells to each other and to things outside the cell, as cells move and communicate.

Turning mRNA into proteins involves a process of *translation*, so called because proteins are made up of amino acids, which are an entirely different chemical beast than nucleic acids. While DNA and RNA are composed of four molecular units that bond in regimented ways to form ribbons or helices, proteins use twenty amino acids, each very different from the others. In contrast with the sameness of the bases that form the structure of DNA's double helix, amino acids are something like a variety of Lego pieces that can be used to assemble an awe-inspiring range of structures and shapes, depending on the pieces used and the order of assembly. The order depends on the translation of a language of four letters into one with twenty letters.

As there is no rule to the order of the letters in DNA, there's no rule to the order of the letters in RNA, other than that it has to be a copy of the message in the DNA, with U substituted for T. Any chemical base can be followed by any other chemical base. So, using simple arithmetic, we can calculate that translating As, Cs, Ts, and Gs into all of the naturally occurring amino acids will require combinations of three letters. This is because copying one letter only per amino acid would yield only four amino acid codes, and copying combinations of two letters per amino acid would yield only sixteen (4×4)—less than the twenty amino acids that make up proteins. If, on the other hand, the code for an amino acid is three letters long, you'd have sixty-four ($4 \times 4 \times 4$) options—more than enough for the task.

A number of experiments have demonstrated that triplets of bases do indeed correspond to the amino acids, with some amino acids coded by more than one triplet. There is also one triplet— ATG—that indicates where the genetic code for the amino acids making up a protein starts, and three triplets—TAA, TAG, and

TGA—indicate where the code should stop being translated. This code applies to all living animals and plants; it is universal, suggesting the breathtaking possibility that all DNA descends from one successful act of molecular invention eons ago. The translation process is carried out by amazing molecular machines called ribosomes, made up of proteins and RNA, that scan the mRNA ribbon as if it were a telegraph tape and assemble proteins according to what they read.

Now we can see, in general terms, what a gene is and how it translates into an observable trait. DNA is unzipped so that sections can be transcribed into RNA, which in turn is translated into, say, an enzyme that will perform a job within the cell. The places on the chromosome where the transcription of DNA into RNA starts and ends are also marked by specific runs of As, Gs, Cs, and Ts, and it is these intervals that we call *genes*. So, genes gain "meaning" in terms of the RNA and proteins that arise from copies and translations carried out by the readers, messengers, and translators. This

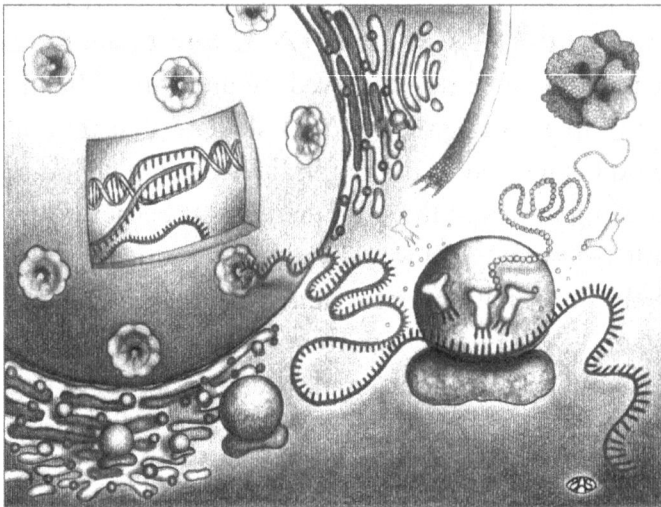

FIGURE 2. The central dogma of molecular biology states that the DNA double helix is replicated within the vault that is the nucleus, where it is also transcribed into messenger RNAs. Some of these are sent to the cytoplasm, where they are translated into proteins within nanomachines called ribosomes, represented by the large structure moving along the mRNA.

raises questions about the detailed workings of these mechanisms and how these related processes started, but we'll have to leave that for another book.

Trawling through the human genome in search of the instructions for proteins is like getting an instruction manual that is about six billion letters long and having to find about twenty thousand words interspersed throughout the text that tell you how to assemble your bookshelf. Through trial and error over the years, we have learned to identify those messages, locate them on chromosomes, and, more often than not, decode them. However, the order of the instructions is not obvious, and if we wanted to build an organism from the sequence of letters in the DNA, we would need to try each and every combination of letters and watch what happens.

As an instruction manual, the genome is not particularly helpful. As a book, it is difficult to read. Even so, it contains information about the pieces, the tools, and the materials that, somehow, will come together in the form of an animal or plant, you or me. But even if we can find some code for specific traits deep in the genome, it remains unclear how those messages in the DNA are transformed into the complicated tissues and organs that our lives depend on. The solution to this puzzle begins to unfold when we look more deeply at the meaning and expression of individual genes.

FUNCTION AND DYSFUNCTION

As we saw above, before Mendel, there were hints that disease was inherited, but what exactly was inherited—the material cause of the disease that was passed on—was not clear. The first disease discovered to be heritable in a Mendelian fashion involves the instructions for an enzyme, homogentisate oxidase. In alkaptonuria, a person's body lacks enough of this enzyme to break down a chemical called homogentisic acid into its component parts, so the acid builds up in the urine. The buildup leads to arthritis and other problems. When exposed to air, the acid turns black; thus the condition is commonly called "black urine disease." This rare phenotype got the attention of Dr. Archibald Garrod at Great Ormond Street Hospital in London in the last years of the nineteenth century.

Garrod started to keep track of infants born at the hospital whose diapers were soiled with black urine and found that the condition ran in families. And he found that it ran not just in families but in the children of first cousins. Garrod consulted with William Bateson at Cambridge, who introduced him to Mendel's experiments with peas. Because alkaptonuria was very rare, it was clearly a recessive trait. In 1902 Garrod published a paper declaring that children inherit "chemical individuality"; after the word *gene* gained currency, alkaptonuria became classified as a genetic disease resulting from a trait or deficit that was inherited, a *mutation* in the gene with the code for homogentisate oxidase. Once this was known, the disease was easily cured by providing the missing protein.

Because alkaptonuria is caused by lack of an enzyme due to a mutation in a gene, scientists thought that genes were always associated with enzymes. But as it happens, many genes aren't linked to a trait in such a simple manner, and chasing them leads to surprises.

Around 1900, a retired schoolteacher called Abbie Lathrop set up a business breeding animals to sell as pets in Massachusetts. Prudently, she chose to start with mice and rats—small animals that reproduce quickly, which would allow her to turn some early profits. One of the first pairs of mice in her menagerie were Japanese waltzing mice, first bred as pets in seventeenth-century Japan and China. This mouse strain exhibits distinctly unusual behavior: rather than sitting still or running in straight lines, a waltzing mouse will run in circles, spinning on an axis in a manner that resembles dancing. Sometimes it won't even run; it will simply pivot around a hind leg, over and over, hundreds of times. Its head will bob around in circles too. You might see why these mice were bred, appealing as they would to humans' curiosity in the days before TV and the Internet.

Over time, the offspring of those mice and other "fancy" mouse strains that Lathrop bred totaled more than ten thousand. She knew that these fancy traits were hereditary and that she had to breed waltzing mice with other waltzing mice to ensure their off-spring were waltzing mice too. So her generations of mice were inbred, mouse sibling bred with mouse sibling, mouse cousin bred with mouse cousin. Within a few years, however, she noticed that some of her inbred strains were developing lumps on their skin and that their offspring had lumps too. The lumps looked like tumors to her, and she contacted some scientists for their opinion. Eventually she ended up working with pathologist Leo Loeb at Washington University in St. Louis to identify inbred strains of mice that were more susceptible to certain cancers. From 1913 to 1918 they published ten pioneering papers on cancer, suggesting a degree of inheritance in mice, establishing that the tendency to mammary tumors, in mice, was hereditary. With this work, the mouse entered the lab as a model for cancer biology and has never left. Nor did the idea that genes cause diseases, particularly cancer.

It was one thing to say that the colors and textures of peas are inherited, but quite another to say that neurological conditions and cancer are. Remember, this was more than three decades before Avery had observed the transformative power of DNA and nearly half a century before Watson and Crick identified the chemical structure of DNA in the string composed of As, Cs, Gs, and Ts.

Today, when we talk about mutations, we refer to alterations in a section of the text of DNA. This may amount to substitution or deletion of one or more letters, which has an effect on specific traits and is transmitted through generations. There are several kinds of changes: a single letter can be swapped with any of the other three, a blank can be introduced, or a passage of code may get repeated. Any change in a stretch of letters associated with a gene creates an *allele*, a new form of the gene. If you imagine the "word" of a gene as a verb, then the alleles of the gene are like different tenses, which change the verb's meaning. When the change in the gene is transcribed into RNA and translated into a protein, these changes determine what the protein can and can't do. If the protein is needed for a particular body function—like breaking down homogentisic acid into its component parts—altering the original text leads to a deficiency in functions of the organism. Such a deficiency might be as simple as not having enough of an enzyme that creates smooth skin, so that you end up with wrinkled peas.

A simple example will illustrate this. Let's take a sentence made up of three-letter words, just like the triplets that code for amino acids: THE CAT ATE THE RAT AND WAS ILL. Deletion of the A in CAT destroys the meaning of the sentence, as it would now read (remember we have to read in triplets) THE CTA TET HER ATA NDW ASI LL: nonsense. Thus, a mutation of this kind would destroy the function of the coded protein. On the other hand, a slight change of a letter in one of the triplets would not destroy the sentence, and we might still get its meaning, depending on where the change happens. If the A in CAT changes to an E, THE CET ATE THE RAT AND WAS ILL still makes sense, but if the T changes to a W, we will not know what animal ate the rat as the triplet would read CAW instead of CAT. Such a change could have serious consequences for the function of a protein.

What Garrod associated with alkaptonuria and Lathrop was breeding in her fancy mice were mutations. In fact, the reason for the waltzing mouse is a mutation in a gene encoding a protein that keeps together the cells in the inner ear that control balance. The same genetic mutation in humans causes Usher syndrome, which causes balance, hearing, and vision problems. The protein is not an enzyme, and herein lies the great challenge of genetics: how to work out the normal function coded in a gene as compared to the dysfunction that results from a mutation when the gene's product is something other than an enzyme. The problem was first brought into focus by mutations that affect the tail of mice.

T IS FOR TAIL

In the 1920s, Nadine Dobrovolskaya-Zavadskaya, an émigré who left Russia after the Revolution and was working at the radium institute in Paris, became interested in the possibility that radiation caused mutations. The reason for this was the stream of deaths, most of them from anemia, bone fractures, and tumors, in women who had been using the fluorescent properties of radium to decorate a number of household items.

In a collaboration with the Institut Pasteur, Dobrovolskaya-Zavadskaya irradiated the testicles of male mice to see if, after breeding, this would create mutations in their offspring. After three thousand crosses like this, she found two *mutated strains*—that is, mutations that stayed true over multiple generations. One of these had very short tails. She called this mutant *T*, for tail (now called *Brachyury*, which means *short tail* in Greek). Following the convention among geneticists, she used a capital *T* to indicate that this was a dominant trait, inherited by offspring even when only one parent had it. And, as it happened, the name of the mutant was transferred to the gene associated with the mutation.

Dobrovolskaya-Zavadskaya had uncovered a mutation that behaved in a very particular way. When one copy of the *Brachyury* gene had been destroyed due to irradiation of the hereditary line's "patriarch," the offspring were born with a tail, but the tail was short. Remarkably, if both copies of *Brachyury* were missing, the offspring died as an embryo in the womb. Hers was a phenomenal piece of research. It would not be the only time a study designed to learn about cancer instead provided lessons in development—that is, the way in which organisms are built. The dead mice hinted at something tantalizing hidden behind the mutations.

At the time, Dobrovolskaya-Zavadskaya did not believe radiation was causing changes in mouse genes. Instead, she thought radiation was "a revealer of a pre-existing latent condition," that it destroyed something that had kept a mutation in check. Later scientists decided to look at what was left in the womb in cases where the mutation was lethal. They found embryos with short spinal cords, confused chest muscles, and no tail at all. This suggested that if *Brachyury* was associated with an enzyme, then the defects caused by its loss were occurring along a spectrum. Less *Brachyury*, less function, more defects—the stronger the phenotype, or observable physical manifestation of the genotype.

It is easy to understand the meaning of genes that provide instructions for specific enzymes as in alkaptonuria. In this case, the
enzyme has a clear function: to break down a substance, homogentisic acid, into its component amino acids so the body can use them.
In its absence the substance accumulates and causes the disease;
we can overcome a deficiency by providing the missing chemical.
This is how enzyme-replacement therapies in medicine work to
treat some diseases. The genes behind mutations such as *Brachyury*
presented a serious riddle, however. Whatever this gene coded for,
it appeared to define not just the length of a tail but the number
and shape of vertebrae and muscles—perhaps, the very organization of the body. While it is relatively easy to understand the consequences of the mutation, it's difficult to understand what the
normal version of a gene codes for or how it contributes to making
a normal pattern of vertebrae.

HERE BE DRAGONS

Breeding *Brachyury* mutants suggested that alterations in genes
could cause disruptions in the development of an organism. But
was *Brachyury* special? Farmers had known for a long time about
sheep with hereditary cyclopia (eyes fused into one in the middle of the forehead) and others with hemimelia (loss of parts of
a limb) and polydactyly or syndactyly (gain or fusion of digits).
The inheritance of these traits suggested that these aberrations
might be associated with mutations and could be associated with
genes. If so, what would the proteins encoded by those genes look
like? What were they doing? More intriguingly, these monsters
suggested that there was a universe of monsters waiting to be discovered and that, from them, one might learn how animals are
put together. To find them, one would have to do a large-scale,
organized mutant hunt.

As fast as mice breed, they could not reproduce quickly enough
to help scientists sort out the relationship between genotypes and
phenotypes and certainly not to illuminate how to create muta-

tions in a systematic way. Such a project demanded a fast breeder that leaves lots of progeny and does not occupy much space. Most especially, if you're interested in observing what happens in lethal mutations, one needs an organism that develops outside its mother.

Enter the "lover of dew," *Drosophila melanogaster*, the fruit fly, which soon became the star of genetics research. It reproduces rapidly, progressing from fertilized egg to fertile adult in just ten days, not to mention copiously—a well-fed female will lay one hundred eggs each day. Like many other insects, including butterflies and moths, it has two lives: first as a maggot, which then metamorphoses into an adult fly with a pair of wings and three pairs of legs. Both the maggot and the adult fly have segmented bodies, with each segment exhibiting particular characteristics, which come in handy when collecting mutants and defects.

Drosophila was not chosen for its flashy looks. It has an unremarkable appearance—unless, that is, you look closely, as famed geneticist Curt Stern did. "When I look at *Drosophila* under the microscope," he wrote, "I marvel at . . . the head with giant red eyes, the antennae and elaborate mouth parts; at the arch of the sturdy thorax bearing a pair of beautifully iridescent, transparent wings and three pairs of legs." Instead, *Drosophila* has been a wonderful partner in the study of genetics because, in addition to being small and short-lived, it has only four chromosomes. This makes the job of the scientist trying to induce mutations in genes and map them to chromosomes that much easier. There's less text to tinker with.

Starting in 1910, a small group of American scientists led by Thomas Hunt Morgan unraveled Mendelian inheritance patterns by turning to *Drosophila*. My choice of the word *patterns* is intentional here, because *Drosophila* is a creature of patterns. Its wings are sculpted with what are called "veins," because they look like the veins that carry our blood, though they're nothing of the kind. These veins crisscross the span of the wings in a precise pattern that is the same from one fly to another. The "sturdy thorax," or midsection of the insect body, which so impressed Stern, is decorated with bristles, arranged in a precise order that is the same from one fly to another too. The regularity of these patterns made it easy for Morgan and his colleagues to find flies with unusual deviations that were passed on from one generation to another and to map these to changes on the chromosome. By 1927, Morgan's research group was able to confirm that the shape of the fly wings, the presence of bristles in the thorax, and a host of other traits obeyed Mendelian inheritance. They also learned that traits mapped to mutations and these to genes that were arranged in a fixed order on the chromosomes, with this order transmitted across generations.

Over the years, Morgan's group, along with other labs, created an enormous catalogue of fly mutants. There were mutants where the eye color changed from the normal red to white, cinnabar, brown, or ruby. There were mutants whose bristles were shorter, thinner, stubbier, or more or less dense, or whose eyes were smaller than normal, or whose thorax was a different color. And sometimes a monster would appear. One curious specimen, first discovered in 1915, looked as if it were trying to grow an extra pair of wings, and sometimes an extra pair of legs, behind the normal ones. A more disturbing mutant had a leg protruding from its head. Both of these flies could breed and pass on these significant defects to their offspring. The mutations were associated with genes.

The four-winged mutant fly caught the imagination of Edward B. Lewis, a young geneticist at the California Institute of Technology in Pasadena. Lewis had studied fruit flies as an undergraduate and had even bought them as pets when he was a child. He was also interested in how mutations could be provoked and transmitted, having studied the medical histories of survivors of the atomic bombing of Hiroshima and Nagasaki. During the 1950s and 1960s, he spent years patiently breeding *Drosophila*, fixated on the mutants that developed an extra pair of wings, looking closely at their bodies and teasing out the similarities and differences in their chromosomes. The job of a geneticist is to make good the saying "The devil is in the details." No matter how the duplicated wings formed, Lewis realized that every single specimen had changes in a particular region of chromosome 3.

One small difference he noticed was in the *halteres*, two short, clublike sensory organs that some insects use to help guide their flying movement. In *Drosophila*, the wings are usually attached to the second segment of the thorax, while the halteres are attached to the third segment, which is small and insignificant. But whenever he observed four wings, the extra pair always appeared instead of the halteres, and the small third segment became bigger. He could see that in many of the four-winged flies, the second set of wings were actually halteres that had partially transformed into wings. Tinkering with the mutants, he engineered a magnificent, fully fledged four-winged fly in which the third segment of the thorax was replaced by a complete copy of the second segment, with a perfect second set of wings. Appropriately, he called this fly a *bithorax* mutant.

In his *Drosophila* crosses, Lewis discovered other mutants, including flies where the position of the abdominal segments was swapped. These swaps had a pattern: the mutation acted on a segment to make it look like a segment that would be nearer to the head in a normal, nonmutant fly. So the segments in the abdomen could be transformed to have the features of the segments in the thorax, which is closer to the head, but segments in the thorax couldn't be transformed to have features of the abdomen. For example, *Drosophila* have legs on their last thoracic segment, and in some of Lewis's mutants, they acquired legs on the first segment of their abdomen. As Lewis mapped the mutants' genes, he saw that the genes with changes were all very close together in the same segment of the fly's third chromosome, so he called the ensemble of genes the *bithorax complex*, in keeping with the name of the original mutant. Another group, led by Thomas Kaufman at Indiana University, observed a mutant with a leg coming out of its head (right) instead of an antenna, called *Antennapedia*. They noticed that this mutant was the first in a small collection in which the front part of the body was altered in some way. This group of genes came to be known as the *Antennapedia complex*. Together, the bithorax and Antennapedia mutants spanned the length of the fly in an

FIGURE 3. A normal fruit fly, *Drosophila melanogaster*, has just one pair of wings and six legs (*left*). In contrast, *bithorax* mutants exhibit a duplication of the middle body segment and bear two sets of wings (*center*), while *Antennapedia* mutants have extra legs coming out of the head (*right*).

orderly manner. Mutants in which body parts get lost, mixed up, or repeated are called *homeotic*, because they hold clues to *homeosis*, or the structural development of organisms.

When Lewis published his discovery in 1978, he said that *bithorax* mutants showed that genes were involved in defining the structure and appearance of regions of the fly body.[2] Together the genes associated with Antennapedia and bithorax spanned and shaped the length of the body of the fruit fly.

By the early 1980s, a small menagerie of monsters had been assembled through genetic studies. Some were strange. The mutant *Krüppel* (*cripple* in German) led to flies lacking most of their thorax; *bicaudal* (*two-tailed*) produced a maggot with a tail in place of the head. Though these mutants did not live long, like *bithorax* and *Antennapedia*, they seemed to hold some key to unlocking the secrets of how organisms are made.

In the 1970s, in the Laboratory of Molecular Biology in Cambridge, England, the South African geneticist Sydney Brenner was exploring how the nervous system is made. He picked an even simpler organism than *Drosophila* for his investigations: the round-worm *Caenorhabditis elegans*. This small, easy-to-grow organism's development from egg to worm can be easily followed under the microscope. The "elegans" part of the name comes from the undulating, elegant movement the worm makes as it searches for food. Brenner poked the worms, observing whether they lost their graceful movements or moved only to the right or the left or not at all in response, and decided this was a good test of how the nervous system was organized. Then he made hundreds of mutants and poked them to find those that did not respond or did the wrong thing. He was on his way to identifying genes involved in the building and function of the nervous system.

In the late 1970s, inspired by this work, Christiane Nüsslein-Volhard and Eric Wieschaus, then working at the European Molecular Biology Laboratory in Heidelberg, devised an experiment to learn how a fly is made: they created mutants whose eggs did not hatch and looked to see which genes were associated with the lethal outcomes.

If everything goes normally, twenty-four hours after a *Drosophila* egg is fertilized, a little maggot, about one millimeter long, emerges. There's a head at one end and a tail at the other. In between there are eleven visible segments, each with a distinct and precise pattern. The whole body is protected by a waterproof layer, the *cuticle*, which has a pattern sculpted into it. If development has failed, the egg rots before a maggot has formed, or the maggot dies in the egg. In the rotting eggs and dead maggots, a ghost of the embryo remains ready for forensic analysis. Nüsslein-Volhard and Wieschaus would use this as a reference for how well development had gone.

As a scientific experiment, this hunt for mutations had its difficulties. Nüsslein-Volhard and Wieschaus, now joined by Gerd Jürgens, induced mutations and set up thirty thousand crosses. Keeping track of the progeny of these crosses and cataloguing the mutants accurately was time-consuming, painstaking work. And although they supposed they would be able to see what had caused the embryo's development to fail, it might not be obvious. Worse, each dead maggot might have died for a unique reason, with no clear pattern linking mutation with mortality. But, looking closely, they spotted patterns. Mutations could be clustered into categories, or classes. There was method in the madness of their monsters.[3]

One class of lethal mutants had two tails instead of one head and one tail, reminiscent of the *bicodal* mutant, above. In some mutants, the body segments between the head and the tail were missing; in others the patterns of each segment seemed to have been laid down in reverse. Mysteriously, some were missing either only odd- or even-numbered segments, and some had two bellies. In some dead maggots, the cuticle was missing.

Nüsslein-Volhard, Wieschaus, and Jürgens followed the tradition by which mutants are named by their distinctive phenotypes, injecting a sense of playfulness into their gallery of monsters: *Toll, snail, hunchback, hedgehog, even skipped, odd skipped, knirps, crumbs, bazooka* . . . the list goes on and on. In some cases, a lethal mutation was related to a previously identified *Drosophila* gene that could produce viable offspring. A mutant in which all segments were fused together happened to be a version of the gene *wingless*, where flies developed without wings. One in which the belly side of the cuticle was missing was an allele of *Notch*, where flies have notches in their wings. This was curious. Much as Lewis had imagined, the discoveries seemed to prove that genes were capable of performing a variety of functions.

Young scientists flocked to Cambridge to work with Brenner, and to Tübingen, Germany, where Nüsslein-Volhard had moved, and to Princeton, New Jersey, where Wieschaus set up his lab, to hunt for more genes linked to development. The fly and the worm were gifts that kept on giving, but a bigger prize would go to researchers who could uncover how bigger animals are made. An American scientist, George Streissinger, had shown that a household pet, the zebrafish (*Danio rerio*), might be a good specimen for this research. He had identified some recessive mutations that created mutant monsters that might be examined to understand which genes are necessary for development. And although much larger than a fly or a worm, the zebrafish breeds quickly, with no more than three months elapsing from fertilized egg to reproductively mature adult. Better yet, it has a transparent body, allowing researchers to see the fish's organs as it develops from larva to fry to adult.

In a heroic effort, Christiane Nüsslein-Volhard set up an ambitious screen for zebrafish mutants similar to the ones she had carried out in *Drosophila*. The experiment presented a logistical challenge of nearly military proportions. You can breed and track flies in small bottles placed in an incubator the size of a fridge. With fish, you need to maintain aquariums—seven thousand of them to generate the four thousand families of fish that she followed for a few years, with four crosses per family. She and her team collected 1,163 zebrafish mutants associated with 369 genes. One of her former students, Wolfgang Driever, set up a parallel screen, identifying 577 mutants and 220 genes. Again, the mutations could be organized into classes and affected the processes of organization of the body, the placement and structure of organs and tissues, and the patterns of colors in the skin.

Finally, it was time to do similar screens with mice. Developmental biologist Kathryn Anderson, who had participated in the characterization of gastrulation mutants in *Drosophila*, set up small, continuous screens at the Sloan Kettering Institute in New York that are still ongoing; remember the mouse needs space and time.

Wherever one looked, lab-created mutants were saying it was possible to disrupt the development of an organism—seemingly any organism—by inducing changes in genes. It seemed obvious that inherited malformations observed in the natural world probably arose in the same way. Still, the perennial, and often forgotten, problem of genetics remained. Scientists knew what happened when they disrupted a function, but this did not contain many clues about the actual function of what was being disrupted. One way forward was to dive into the genome, find the gene, and, using the genetic code, try to decipher the proteins encoded in that stretch of DNA. Maybe then we could understand not only the making of an organism but the relationship between genes and complex traits like an eye, a leg, or the activity of a neuron.

COMMON BASES

In the 1980s, as a graduate student at the University of Chicago, I was bewitched by mutations like *Brachyury*, *bithorax*, and *bicaudal* and got the news of the Heidelberg screen. Could these monsters tell us how organisms are put together? This question led me down the path that has marked my career. What magical proteins were encoded in those genes? If unique combinations of enzymes produced different forms of life, science would need to learn how they work. But what if there was something else? The only way to know was to fish out the DNA associated with genes and see what proteins it encoded.

Several research groups were looking for answers along these lines. In California, Ed Lewis had teamed up with biochemist David Hogness to look for the DNA associated with the bithorax complex. Two other teams, one in Indiana headed by Thomas Kaufman and the other in Basel led by Walter Gehring, were focused on studying the DNA related to the Antennapedia complex. It was hard work, but several years of trawling through chromosomes, using techniques developed for this purpose, laid bare for examination specific pieces of DNA linked to both mutations. There were three stretches of DNA coding for RNA messages in the bithorax complex and seven in the Antennapedia complex. Yet, whatever the message hidden in those genes, it was not being translated into enzymes. Then, late on a summer afternoon in 1983, at a meeting of European scientists doing research on *Drosophila* genetics in Cambridge (UK), some members of Gehring's lab shared an odd observation: a fragment of DNA, a small part of the text, appeared to be the same in several of the genes in both complexes. It was 180 letters long and had the code for sixty amino acids. This code came to be called the *homeobox*, because it contained duplicates of homeotic genes first seen in *Drosophila*. The fly genes were copies of each other.

Researchers in Gehring's lab had the insight that homeobox DNA could be used as a bait to fish for similar genes in other animals. Because the As only ever pair with Ts, and Gs only ever pair with Cs, you could use an identified strand of DNA to fish for a complementary piece of DNA in another genome. Scientists collected DNA from a whole bunch of worms and insects, and, lo and behold, the homeobox was everywhere. A lab nearby decided to check its frogs and found that they also had the genes. As Lewis put it, the so-called *Hox* genes were "a flying carpet" that allowed scientists to locate genes across many animals, from flies to higher vertebrates.

It was an exciting time to be a biologist. I remember the thrill of some new discovery being reported on an almost weekly basis. The expression of *Hox* genes was restricted to parts of the body in mutants. Combinations of *Hox* gene expression defined individual parts of the body. Most surprisingly, the genes were arranged along the chromosome in the order they were required in the organism. When it came time to look at their RNAs and proteins, the order held up.

Today we know that these *Hox* genes are present in all organisms, including humans, and that their order in the chromosome follows the rule of the fly. Their pattern is a universal outline, or map, of the organism in the genome. This startling finding brought new meaning to the words of William Blake:

> *Am not I*
> *A fly like thee?*
> *Or are not thou*
> *A man like me?*

While a fly and a mouse don't have much in common by way of looks, the commonality of *Hox* genes—their *conservation* across species, as scientists put it—suggests a common genetic legacy.

More to the point, the proteins coded in the *Hox* genes were absolutely not enzymes. Later it was found that Hox proteins are what we call *transcription factors*, proteins that bind to DNA to activate genes, making the homeobox crucial to this task. This suggested that *Hox* genes are tools that contribute to the pattern of the body in the manner that Lewis had imagined: controlling the expression of other genes in specific regions of the chromosomes and using the products of those genes to mark the differences between parts of the body. Some of these genes code for other tools, or transcription factors, but others code for products that are building materials: cytoskeletal elements, adhesive proteins, or components of the extracellular space.

Soon the techniques used to fish for *Hox* genes were being used to look for genes across species, and the genes associated with *Drosophila* mutants were found in worms, fish, and mice. This too was shocking. The "words" associated with building different organisms were the same. Vertebrates have additional words that insects do not have, but for the most part, the messages in the DNA are the same in most animals. Again, many of the proteins coded in the DNA are transcription factors, and several are related to the homeobox, and this suggests that development is about controlling gene expression. But there are also proteins associated with components that make up individual cells and some associated with metabolism. In many instances, the genes revealed in *Drosophila* screens have been associated with severe disease in humans—for example, *hedgehog* genes are mutated in human basal cell carcinomas. By the end of the 1990s, we had to accept that, from a genetic perspective, we humans are not as special as we might once have thought.

HEARTLESS, EYELESS, AND OTHER ODD HAPPENINGS

Behind the conservation in the toolbox lurked another, bigger surprise. It is not difficult to see the multitude of differences between animals. Some are obvious: the long neck of the giraffe, the trunk of the elephant, or the way our forelimbs are paws, wings, or hands. Others are hidden from our view: invertebrates pump blood and breathe air like we do, but the organs that perform these tasks are very different in appearance from ours—a pipe with branching tubes or gills versus ventilating lungs. In another example, *Drosoph-*

ila have seven hundred or so separate visual receptors in each "eye," compared to our two mammalian eye complexes. Vertebrates have some cells, like the insulin-producing cells of the pancreas, that invertebrates simply lack. But when we started to map genes to body parts and to function, something astonishing came up.

Take the case of a gene called *tinman* found in *Drosophila* and other insects. This gene is associated with the development of the fly's simple heart in the form of a tube that acts as a pump—in *tinman* mutants, as in the Tin Man from *The Wizard of Oz*, the heart is missing (geneticists are playful with names)—and the gene's RNA and protein are expressed in the cells that will form the tube. A similar gene is found in fish, mice, and people, and in all cases, it is associated with the early stages of heart development, which also involves the formation of a tube. In mice and humans, it carries the less evocative name *Nkx2.5* and, as in the fly, is associated with the early stages of the building of the heart. The gene codes for a transcription factor, a tool like the *Hox* genes, to reach out to other tools associated with the making of the cardiac tube.

This relationship between our heart and the heart of an insect at the level of genes is very surprising, but it is not an isolated case. Other genes associated with heart development pop up in a range of organisms, as *tinman/Nkx2.5* does. This suggests that we can trace a line of ancestry from the insect heart to our own, in terms of not just the organ's structure—a tube that becomes looped and acts as a pump—but also the genes associated with how this happens. Similarly, the same genes are required to build the branching trachea of insects and the branching bronchi and bronchioles that characterize the lungs of mice and humans. Even *Brachyury*, the gene that causes short tails in mice, can be found in both insects and humans, and it is always associated with the development of the back end of the animal. Similar organs in different animals are underpinned by similar genes.

Today, labs can sequence genomes at an amazing pace; a human genome, for instance, can be sequenced in a single day. Machine learning also makes it possible to compare the vast libraries of base pairs in different genomes to find sequences, or "words," that are similar or the same across organisms. Using this technology, we have learned that the same or very similar genes exist in simple animals like sponges and jellyfish and complex ones like us. This makes some sense, given that there is a common origin to life on earth.

But it's vital that we don't confuse mutants with the genes or these with the function of the proteins they encode. The *tinman* and *Nkx2.5* genes do not make a heart; they are associated with the construction of a heart, but it is not clear how. The one thing these observations tell us is that it's not only the genes that are conserved but also their functions, whatever these are. The fruit fly mutant *eyeless* exemplifies this convergence. Flies with the *eyeless* mutation typically lack the usual array of visual receptors and structure of a fly eye. When geneticists tinker with the *eyeless* gene and place its RNA in another part of the fly's body to promote the expression of the protein it encodes, an eye will appear wherever the RNA coding for *eyeless* has been placed. *Eyeless* "makes" eyes. That's very interesting, but it's only the beginning.

It turns out that a similar gene exists in humans, where it is called *PAX6*. A mutation in *PAX6* can result in aniridia, where the iris doesn't form fully or at all. Other congenital abnormalities of the eye are related to *PAX6* too. Scientists in Walter Gehring's lab decided to see what would happen if the human *PAX6* gene was inserted and expressed in a fly. In this experiment, an eye appears wherever the *PAX6* gene is expressed—exactly what happens when *eyeless* is expressed. Yet, strangely, instead of humanlike eyes consisting of one large visual receptor, flies with *PAX6* develop fly eyes, with seven hundred or so receptors.[4] The human gene placed in a fly still makes an eye, except that now it is a fly eye. The same has been found to occur with other fly genes, like *Dichaete*, which is involved in the development of the nervous system in the fly and has a homologue in mice and humans called *Sox2*. Even the famed *Hox* genes follow this pattern. Human and mouse *Hox* genes will mask over defects caused by mutations in fly *Hox* genes and contribute to making a fly's body.

What an odd book is this book of life, where a word in one language makes the same amount of sense in the grammar of every other language.

Over the past century, biologists, including at one time myself, have sought to explain what makes animals different from each other, weaving a story out of a dazzling parade of genetic studies. This started with a search for differences in genetic mutants and ended with the finding that the genes revealed by those mutants are universally conserved and that the same genes are associated with the heart of a fly and that of a human. There have also been attempts to tease out the origin of our differences in other features of the genome, like the number of genes that each species has. We have a few more genes than a fly—*Drosophila* has around fifteen thousand genes, while we humans have between twenty and twenty-five thousand—but after we account for the similarities, the differences are not so enlightening.

Then, what's left? We can resort to exploring how genes are deployed across space and time. Perhaps the DNA sequences that provide instructions for when and where genes can be expressed hold the key to the differences between animals. Or perhaps it's the combinations of proteins coded in the genes that matter. This would at least explain why a human gene will make a fly part in a fly, because in the fly, the human gene creates a protein that is part of a collection of proteins, all of which click together, like Lego pieces, to make a fly. There is some truth in all this, but if you sense that something is missing, you are not wrong. Just as a house is not a collection of bricks, an organism is much more than just a collection of genes.

THE LIMITED POWERS OF GENES

If there is a situation in which we can see how much genes matter to how we look, feel, and behave, it is in twins. In particular, identical twins share all of their DNA and therefore present a convenient way of assessing the extent to which genes make us who and what we are. Fraternal twins provide a sort of control in these studies to account for the effects that sharing a womb might have on the development of different traits.

Traits that are similar between identical twins but not fraternal twins are said to be *concordant*, and the degree of concordance offers a way of measuring how much a particular trait is determined genetically. The degree of concordance varies not only depending on the trait under scrutiny but also, surprisingly, from one study to another. This is particularly clear when the study moves away from physical or pathological aspects to intellectual abilities and behaviors. The numbers that result from these studies often confirm our intuitions that much about our body appearance is inherited. After all, don't identical twins look identical? In a more quantitative mode, height shows a very strong concordance, at over 80 percent—but notice that it is not 100 percent. In contrast, many diseases, including cardiovascular disease, have a much lower concordance, in the 20 to 30 percent range. In fact, a study looking at 560 diseases affecting fifty-six thousand pairs of twins included in a database of millions of people concluded that only 40 percent of the diseases had a strong genetic input.[5] Nevertheless, there's a tendency to emphasize high concordances, largely—I think—because we are used to hearing that genes play a large role in our features and chances of developing diseases. Furthermore, nowadays, when we cannot blame genes directly, we talk about *epigenetics*, a system of chemical modifications in DNA and of proteins associated with specific genes that contributes to gene expression by channeling cues from the environment. In particular, epigenetics refers to the influence of individual experiences—diet, exercise, habits—on the control of gene expression. There is likely something in these observations, but often they read like the proverbial can kicked down the road; if we cannot blame genes and only genes for a phenotype, we shall just look for an explanation in something that regulates the genes, even though, in the end, it is always the genes.

The impression of resemblance between identical twins is drawn from our ability to recognize faces, which focuses very much on similarities; this interest of our mind in similarities not only makes them look the same but probably enhances this impression. Then, knowing that they share all their DNA, we draw the conclusion that the genome contains a blueprint for the face. This makes sense, but the same conclusion can be reached if we believe that rather than a blueprint, identical twins share the tools and materials needed to build a face. It's like assembling bookshelves from a store kit: the final products look identical because the parts in the kits are identical and adjusted to fit perfectly. They also share a blueprint, but the blueprint comes from somewhere else, and someone has to interpret it and put the pieces together. This idea that the similarity comes from the genetic kit is supported by the existence of *doppelgangers*, people who look very much alike even though they are not related genetically. In these cases they have likely inherited the same or very similar genes associated with building faces. Using these individuals, we should be able to find out the nature and identity of those tools, and, not surprisingly, a 2022 study is beginning to do this.[6]

Many characteristics would provide a test for how much genes determine the organization of our bodies. Here I will choose heterotaxias. Typically, we each have two eyes, two ears, and two arms, one on each side, symmetrically distributed across a line that runs along the middle of our bodies. We also have some unique organs that lie on one side or the other of this line. The heart is usually on the left side of the body, as are the pancreas and the spleen, whereas the liver is skewed to the right. In people with a heterotaxia, one or more organs form on the wrong side of the body or are missing altogether, with dire consequences for the person's health. Many cases of heterotaxia are hereditary. Researchers have been able to map these cases to genes, but individuals with the same mutation commonly have different degrees of defect in the formation of their organs. This is typically explained away with the observation that different people have different genes and genomes. But even identical twins carrying the exact same mutation can differ greatly in how severely the syndrome manifests in their body.

In one particularly dramatic case, identical triplets were all born with a cleft lip, where the lip has a gap or split because the two halves have failed to join together.[7] Two of the siblings had a cleft lip biased to the right, while the third had a split in the middle. One of the siblings with the cleft lip on the right also had a severe cleft palate, a gap in the roof of the mouth. They all had the same mutation in their DNA, but the defect had not formed in the same way. When it comes to positioning the mouth with regard to the body's midline, genes can't tell right from left from middle. The simplest conclusion is that there is no gene for positioning organs in the body. As much as we have grown accustomed to talking about how genes determine our lips, ears, limbs, hearts, brains, and even our personalities, if genes can't tell right from left or middle, they simply can't be responsible for doing everything involved in the making of you and me.

The widespread view that genes control so much of our lives and are the architects of our bodies owes much to the success of genetics research over the last sixty years—especially the links established between particular genes and many specific diseases: alkaptonuria, thalassemias, sickle cell anemia, cystic fibrosis, and Huntington's disease, to cite a few you may be familiar with. In many of these cases, this link has led to actual cures, through the supplying of a missing enzyme or, more recently in a small number of cases, repair of the damaged gene. In the midst of these success stories, the treasure trove of mutants affecting the building of organisms has led us to believe, by extension, that the making of an organism works in the same way—that it is primarily the work of genes, with a bit of support from the proteins they code for. This is why we talk freely about genes for eyes, hearts, or hair. In an extreme version of this, we say that some people have a gene for x that others don't—for example, red hair or blue eyes—when we all have the same genes, and we are actually talking about variants, or mutation, of those genes.

But there is a difference between those genes that we can associate with specific diseases and those that we identify when we screen for genetic defects in the building of organisms. In the first case, a piece of hardware at the end of a construction line—a screw or cap for the assembly of the bookshelf—is often damaged, and this is easy to repair. The mutations that affect development are a different matter: often they have to do with design, and at the moment it is not easy to work out how exactly the action of the protein encoded contributes to the process disrupted by the mutant gene.

When you think about it, we have turned the object and method we use to study how we are made, *genetics*, into an explanation and mechanism for how we are made, *genes*. But then you will ask, if genes are not in charge, who is? If the genome is a toolbox, who uses it? Where is the blueprint for our bodies?

The missing character in our story can be gauged from a close reading of the news stories reporting that Craig Venter, the enfant terrible of modern genomics, had in 2010 created a new "synthetic life form." Or had he? Scrutiny of his experiment revealed he'd done nothing of the kind. Instead Venter and his team had replaced the DNA of a very small bacterial cell of the genus *Mycoplasma* with a very small piece of DNA, synthesized in the lab, which contained what he and his team thought was the minimum number of genes needed for the survival of a bacterium. The cell that received this synthesized DNA had not been created anew. The activity of the cell may have changed, because it contained a different set of genes, but the new DNA needed to be in a cell to implement the change. Without a cell, DNA is useless. Saying that this was a new life-form was the equivalent of saying that writing a new computer program results in the physical construction of a new computer, that software can create hardware. In reality, you need a computer to read the program. Software always needs hardware. In the same way, you always need a cell for DNA to do something. A better way to describe what had been accomplished is to say that a cell had been "retooled."

As we shall see in the following chapters, the limitations of the genes as elements of a blueprint become more obvious when we consider the building of multicellular organisms, their development, particularly in the way embryos come together. These structures exist in space, and genes can neither create nor sense physical space. Nobody will deny that genes play a role in the development of an organism—the mutants tell us that—but they are not in charge. What they do, they do under the control of cells. If you were to put DNA in a test tube and wait for it to make an organism, it would never happen. Even if you were to add all the ingredients that allow the reading and expression of the information in DNA—the transcription factors, plus some amino acids, lipids, sugars, and salts to help catalyze chemical reactions—it would still never happen. DNA needs a cell to transform its content into a tangible form. An organ or a tissue, and most certainly an organism, is no more the result of the activity of a collection of genes than a house is an aggregate of bricks and mortar.

To build an organism, as to build a house, you need not only a blueprint but also skilled workers who will interpret the architect's design and assemble the tools and raw materials needed to execute it. In the construction of an organism, cells are the master builders.

two

THE SEED OF ALL THINGS

Robert Hooke was aged just twenty-seven when, in 1662, the newly chartered Royal Society named him as its curator of experiments. A polymath, he had already discovered his eponymous law of elasticity and built a reflecting telescope to peer into the heavens and detect previously unseen stars in the constellation of Orion. He was equally enamored of the hidden world that the earliest microscopes were just beginning to make visible. *Micrographia*, his 1665 book sharing his studies of the minuscule, was arguably the first scientific best seller.

Hooke's drawings caused a sensation. It seemed implausible that a fly's eye could be made up of hundreds of individual visual receptors, but Hooke had seen them under his microscope and drawn them for the educated public with exacting precision. When he turned his eye toward a cork tree, he was equally intrigued by what he saw:

I took a good clear piece of Cork, and with a Pen-knife sharpen'd as keen as a Razor, I cut a piece of it off. . . . I could exceeding plainly perceive it to be all perforated and porous, much like a Honey-comb, but that the pores of it were not regular; yet it was not unlike a Honey-comb in these particulars. . . .

The *Interstitia*, walls (as I may so call them) or partitions of those pores were neer as thin in proportion to their pores, as those thin films of Wax in a Honey-comb (which enclose and constitute the *sexangular cells*) are to theirs.

At this moment in scientific history, the word *cell* made its debut.

These cells, he said, explained the characteristics of cork. More astounding was their number: Hooke estimated that there were more than one thousand in each inch lengthwise. That meant that there were one million in a square inch of cork and twelve million in a cubic inch, "a thing almost incredible, did not our *Microscope* assure us of it." To be sure, what he saw were not actually real cells, as we know them today, but their remnants. He was observing a pattern of packed hexagonal structures arranged as a sheet, generated by what we now know is the cell wall, a ghost of where a cell had been.

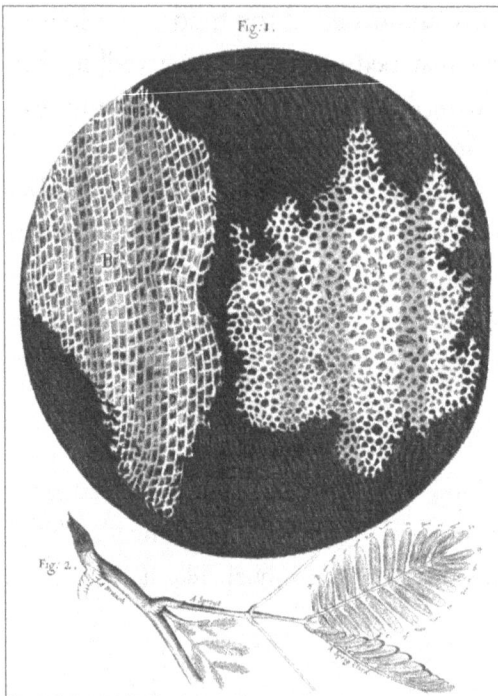

FIGURE 4. Drawing from the structure of cork, showing *cells*. From Robert Hooke, *Micrographia*, 1665.

Cells had gotten their name, and the first real sighting was not far off. A few years later, the Dutch draper Antonie van Leeuwenhoek, using a special microscope he built himself, observed living cells in the form of organisms swimming in drops of pond water. What he called *animalcules*, or little animals, we would today recognize as *protozoa*. He then turned his attention to semen, where he saw thousands of sperm milling around in a dense fluid. He thought these were animalcules too, akin to the organisms he had observed in pond water. The distinction between a cell and an organism was not clear then (and in some cases, it still isn't).

As microscopes got better over the eighteenth and nineteenth centuries, a consensus emerged that all life on earth is made up of cells—not the empty shells that Hooke had described but entities filled with the stuff of life. Thus biology was transformed with what we now call *cell theory*, the notion that cells are the foundation of all living things, the basic units of life. At that time, before genes came into the picture, cells were heralded as the building blocks of organisms. Physicists like to brag about the number of stars and galaxies in the vastness of the universe, but to beat those numbers you need only trade your telescope for a microscope. There might well be ten trillion stars in the universe, but a human being has about forty trillion cells. With eight billion people alive today, there are some 10^{21} (one followed by twenty-one zeros) human cells kicking around on the planet right now. When you add the contributions of other animals, fungi, plants, and single-celled organisms to the panoply, the number of cells living at this moment increases phenomenally.

The devices you're using to read these pages, your eyes, are made up of cells and, among them, a special class of cells that respond to light. Your eyes' photoreceptors share what they sense with other cells in the brain, cells that recognize and interpret patterns in sensations by communicating with each other. The ability to carry out these activities lies not in your DNA but in the arrangement and function of the cells that make up your eyes and your brain. The variety in organization that we observe in plant, fungus, and animal forms—for example, the differences between fly eyes and human eyes, fly brains and human brains—is not found in the DNA, because, as we have seen, genes from flies and humans can be used effectively in each other. Fundamentally, organisms cannot be built up from the utter sameness of the thin chemical strands, with their very specific dimensions and unwavering double helix structure, that comprise DNA molecules and genes. DNA is just a store of information to manufacture RNA and proteins. Although DNA has a role to play in the construction of our bodies, it is cells, not genes, that make you and me what we are.

We are cells, and cells are us, but while we are all familiar with that iconic double helix that resides in their nuclei, cells themselves remain lesser-known entities, support players in the workings of genes. However, it is cells, not genes, that perform your digestion, keep your heart and your brain going, spare you from infections, and allow you to read this book. In contrast with the monotonous structure of DNA, cells exhibit an enormous variety of appearances due to a multifarious internal organization. It is this and the multiple combinations of their component elements that fuel their creative power as agents of form and shape. If we want to understand how cells control genes to make organisms, we need first to look at their inner workings.

CELLS

Many of us can look back to our first glimpse through a microscope. It might have been in a biology class, and the preparation may have entailed a slice of onion, a splash of pond water, even a drop of blood. Whatever the specifics, that microscope was showing us the cell, whether it was a solitary cell, wriggling around and foraging for food, or a group of cells, packed together like bricks in a wall.

Cells come in many different shapes and sizes. It is thought there are at least two hundred different types of cells in the human body, many with individual varieties. In your skin, blocks of densely packed cells, each with the actual appearance of a brick, form a palisade, or wall-like form, called an *epithelium*, that serves as a barrier,

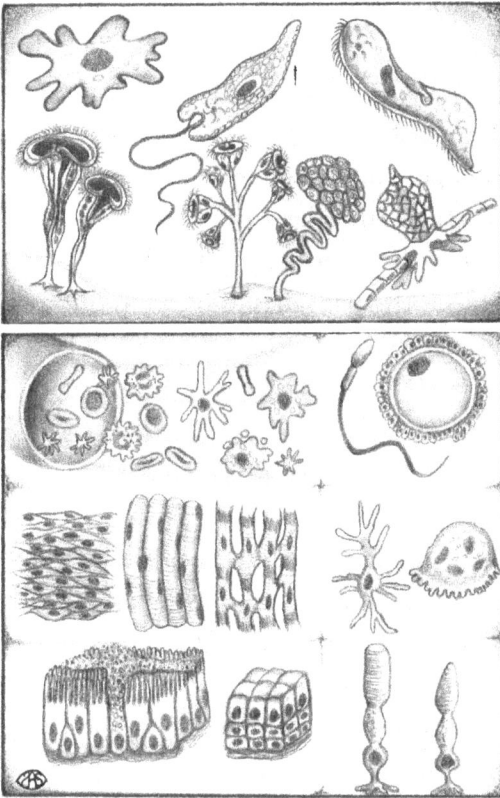

FIGURE 5. Sample of the diversity of cells. *Top:* examples of protista. *Bottom:* cells that can be found in the human body. *From top to bottom and left to right:* blood cells, sperm and egg, different kinds of muscle cells, neurons, epidermal epithelia, and rods and cones from the retina.

protecting your internal organs from damage. Your intestine also consists of an epithelium, in this case a single layer shaped into a cylinder, part of the larger tube that runs from your mouth to your anus, where food is processed and taken up to produce energy for your growth and maintenance. Muscles are reams of special cells organized into elastic fibers that expand and contract to motor your movements and actions. Your heart is an ensemble of a particular type of muscle cell, tightly packed and connected to elements of your nervous system. These cells expand and contract in rhythmic synchrony, beating about one hundred thousand times a day across the whole length of your life, to push blood and the nutrients it carries around your body. Your lungs have a complex branching tree of cells that capture oxygen from the air you breathe and transfer it to your blood cells. Every organ in your body is made up of cells tailored for specific functions: protect, feed, pump, breathe.

The human brain, that seat of our will, our thoughts, and our feelings, is also made up of cells, a very special class of cells that shows the potential of these building blocks in terms of form and function. When, in the nineteenth century, slices of brain tissue were placed under the microscope, scientists saw a puzzle: a twisted, dense mesh of stringlike structures and dense circles. This didn't look like the walled chambers first identified by Hooke, the animalcules of van Leeuwenhoek, or the palisades of our skin or intestines with clear individual components. At the time it was already known that the brain is powered by electricity. Was the structure created by a new and different object, they wondered, that created the miracle of consciousness?

In the late 1880s, a maverick researcher named Santiago Ramón y Cajal, working in Barcelona, set out to methodically stain sections of young brain tissue to reveal what the brain is made of. His methods untangled the complex of knotted strings and led him to realize that he was indeed looking at cells—but cells like no others in the body. Much like other cells, they had a core body, which was crowned by short, spiky extensions, but then each also had a long, twisting cable that branched out into increasingly thinner structures that connected with similar structures branching out from other brain cells. In young brains, the tips of these branching cables appeared to be exploring the environment, looking for information and other cells, looking for partners. Ramón y Cajal was describing *neurons* and their protruding *axons*; the tangle of cables and meshes seen under the microscope was a very dense conglomerate of neurons and axons that dwarfed the cell bodies. It was so dense that it confused the eye. There are more than eighty billion neurons in your brain. On average this is one hundred thousand neurons per cubic centimeter—about the size of a small sugar cube.

As he untangled the mesh, Ramón y Cajal inferred that the brain's electricity must flow through the neurons, starting from the cell body and traveling to the tip of the axon. He was right. Shortly afterward the electricity was shown to be transmitted from one neuron to another neuron where the axons connected, through a

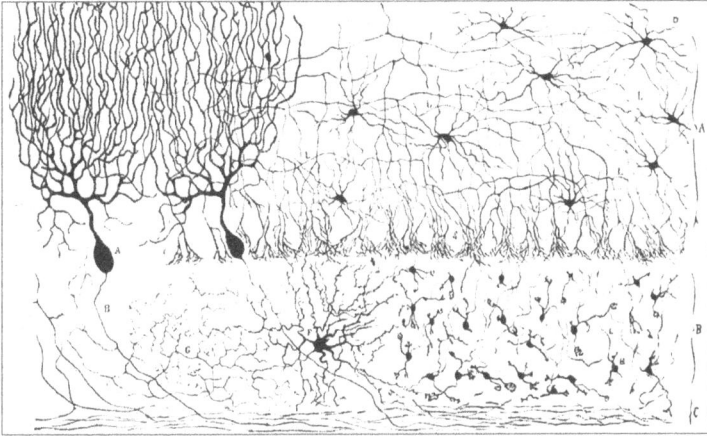

FIGURE 6. Santiago Ramón y Cajal's drawings of different kinds of neurons that can be found in the chick cerebellum. From "Estructura de los centros nerviosos de las aves," 1905.

structure called a *synapse*, from the Greek for *together* and *clasping*. Within the mesh, collections of axons make precise connections with each other. These have been recently revealed in their full glory using modern microscopic techniques. Not only that, but the basic structure of neurons is the same across species, from snails to humans and practically everything in between.

Neurons are a great example of how cells, even cells of the same functional type, are able to take on all sorts of forms and shapes. Ramón y Cajal, an artist at heart, captured in his drawings the dazzling variety of neurons. Some neurons are small and bushy, while others are giants, with big branches resembling gnarled old oak trees. At about four micrometers (0.004 millimeters) in length, the smallest cells in your body are granule cells, the most numerous type of neuron in the brain. It is found in the cerebellum, the "hind" brain common to all vertebrates. The largest cells are motor neurons, which are about one hundred micrometers (0.1 millimeters) in diameter and have axons stretching a meter, reaching from the spinal cord to the tips of your toes. And neurons are found throughout the nervous system, not just in the brain. In your ear, small neurons connect with other, more sophisticated cells with structures that sense air movement. Their sensitivity to air waves creates the electrical impulses that generate what we perceive as hearing.

When one looks at neurons in different organisms, one finds the same basic structure described by Ramón y Cajal, but the detailed shapes are very particular to the organism. For example, the neurons of C. *elegans* lack the branching characteristic of the neurons of insects and vertebrates, but they perform similar functions. In other examples of a variety of structures serving the same function, erythrocytes, the cells that move oxygen around the body, display a large range of shapes and sizes across the animal kingdom, and eggs vary from the 0.5 × 0.15 millimeters of a fruit fly egg to the 15 × 13 centimeters of an ostrich egg. It is not yet clear where this enormous diversity in size and shape comes from. But looking inside any cell reveals some common structural elements to play with. All two hundred or more different types of cells in your body are variations on a theme, a combination of proteins, lipids, and nucleic acids with a dash of sugars and salts.

SKETCHES OF A CELL

Despite this variation in appearance, often tailored to function, all cells share an organization. Every cell has an outer perimeter, the barrier that defines it as a distinct unit. This perimeter is further delimited by what is called the *plasma membrane*, a semipermeable structure composed of two layers woven with lipids and proteins that protect the contents of the cell. This membrane regulates the interactions between the cell interior and its external world, which, often, is another cell. The diverse proteins embedded in its fatty makeup decide which molecules go in and which come out. Some allow cells to stick to other cells or to repel what they don't like. Some are tuned to survey the surrounding environment for signals, allowing cells to talk with other cells. The plasma membrane of some cells is a launching pad for signals, like hormones—for example, in the pancreas, where they secrete insulin—for which other cells have receptors. Channels in the membrane allow water and ions to be pumped in and out, ensuring a favorable environment for everything happening inside the cell. In the case of neurons, the opening and closing of these channels generates and transmits the electrical impulses from the cell body through the axon to other neurons.

A cell is a world in which proteins are major actors. Inside the protective case of the plasma membrane, we find a dense beehive of activity where proteins stand in for worker bees, and instead of honeycomb and wax, there are more membranes, each with a special composition and function (there are great videos on the web where you can marvel at this dynamic organization). These membranes differentiate the *organelles*, organlike structures that each have a particular job to do. Some organelles operate like factories manufacturing proteins and lipids and assembling them for use. Others serve as garbage disposal units that get rid of what the cell does not need anymore. All internal membranes are linked to create a transport system that includes the plasma membrane, serving to ensure that proteins get where they need to be to do their jobs. Floating in this complex space are *mitochondria*, strange cylindrical structures with their own membranes and DNA. Mitochondria provide fuel for the cell. Not surprisingly muscle cells are full of mitochondria. And at the center of all this activity, there is the most familiar organelle—the nucleus—serving as a vault for the chromosomes and DNA. The nucleus has its own membrane, different from the plasma membrane but with a similar gatekeeper function governing the relationship between its contents and the rest of the cell. The membrane of the nucleus links, in ways we still don't understand well, with the system whereby proteins are manufactured, so that when synthesized, messenger RNAs, the copies made from the genes in the DNA that serve as the templates for proteins, find their way into the protein factories.

This system of membranes would be a shapeless, flopping blob if it weren't for a cellular skeleton, or *cytoskeleton*, that provides support, holding together the cell, determining its shape, and defining how and when it moves. The *cytoskeleton* is an array of small

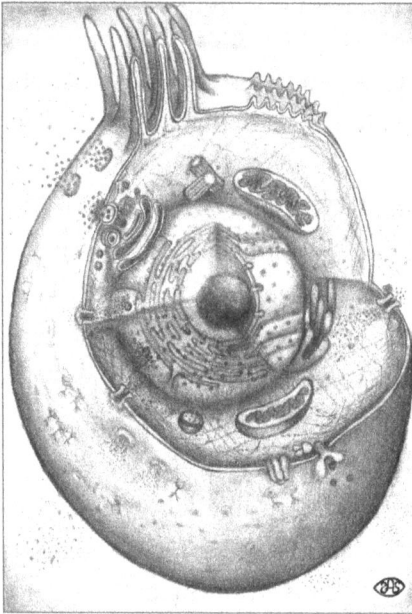

FIGURE 7. Sketch of the structure of a generic cell with the nucleus at the center and various organelles around it within the confines of the plasma membrane.

filaments, made up of the protein Actin, and tubules whose units are made up of another protein, Tubulin, which together give shape to the cell. It is supple and in constant flow; Actin filaments are very dynamic and allow the cell to change shape and adapt to its environment: expand, contract, move. Tubulin creates tracks, more stable than the Actin filaments, that in some instances contribute to the structural stability of the cell and also to its internal highway system. Many cells have a third skeletal component: filaments built of strong proteins, for example, Keratin, that are very stable; your hair is made up of Keratin, and the word may be familiar from the list of ingredients on the side of some shampoo bottles. Unlike our bones or the chassis of a vehicle, the cytoskeleton is flexible and changes all the time in length and spatial organization, which allows cells to adapt their shape to fit alongside their neighbors and within the environment. An active cytoskeleton is the best sign that a cell is alive and well.

This basic organization pervades all cells, from those animalcules that van Leeuwenhoek saw in pond water to the components of an oak tree and the cells in your body. The arrangement and dynamics of those components give identity to the two hundred or so kinds of cells mentioned above. The organization of cells in an epithelium depends on the arrangement of their cytoskeletons, as is the case, but in a different way, for the cells in the blood that survey your body for infections or that provide oxygen to the different organs and tissues and, of course, for the neurons in your brain.

In all their many and varied forms, eukaryotic cells are truly a wonder of nature; they are not just the wooden structures suggested by the observations at the end of the nineteenth century but pliable, adaptive, and morphing entities pulsing with life. But one feature that we take for granted is the secret of their ability to be creative and to conquer the earth. This is their capacity to multiply—a trick that turned the cell you once were in your mother's womb into the trillions that shape your body and make it function, a trick that re-purposes many of the cell's components for reproduction.

DIVIDE AND CONQUER

Organisms have a most important characteristic that sets them apart from inanimate matter like water, mountains, and planets. It is not their ability to grow; the Himalayas and, unfortunately, the seas are growing, and planets form through the aggregation of celestial gases and dust. What sets living organisms apart is their ability to replicate and multiply. This capacity is built into the nature of cells, which grow and, when they reach a certain size, divide in order to multiply. In this way, one becomes two, two becomes four, and the repeated process over time generates the trillions upon trillions that make up you and me.

When the cells of plants and animals multiply, they do so through a process called *mitosis*, where the two so-called daughter cells each have similar volumes, components, and DNA to their parent. Organelles called *centrioles* are vital to this process. These are pairs of short, tubular structures that sit at right angles and close to each other outside the nucleus, surrounded by a cloud of proteins.

Most of the time, they act as the seed for *cilia*, arm- or antenna-like protrusions that survey the chemical environment within and surrounding the cell. When necessary, cilia swirl around to keep the surrounding chemistry well balanced. In some cells, centrioles sow flagella, long protrusions that propel cells' movements; sperm have long flagella that allow them to swim in the uterus in search of an egg. In all cases, the most important function of centrioles is to serve as the foundation stone for tracks of *microtubules*, used to transport materials around the cell, or as anchors holding mitochondria in place.

When it's time for the cell to grow and divide, the centrioles stop what they're doing, leading to the disassembly of the structures they support, move to opposite sides of the cell, and lay down a bridge of parallel microtubules across the cell. The length of these tracks, their distance from the plasma membrane, and the placement of their connection to the centrioles are dictated with the accuracy of a precision engineer. The product of this handiwork is called the *spindle* because of its shape. Once the spindle is in place, remarkable proteins called *dyneins*, which look like giant construction cranes, latch on to the center of a strand of chromosome and slide it toward one of the two poles. This is done methodically, so that each pole gets half of the parent cell's DNA, which will be used as a template to rebuild the whole chromosome. Once the chromosomes are all moved into their new positions, the spindle breaks down. As this happens, the cell is cleaved, right through its middle, with a new membrane forming to divide the two halves. Centrioles go back to other business, and where there was once one cell, there are now two. The form and shape of the daughter cells take after the parent's.

Here we must pause and ask what genes have to do with this complex process. If all cells have the same DNA, where do they encode the shape of two hundred or more different types of cells? True, genes contain the information for the proteins that make up the centrioles, but do they "know" they should only make two? How do genes instruct the centrioles to build the microtubule bridges

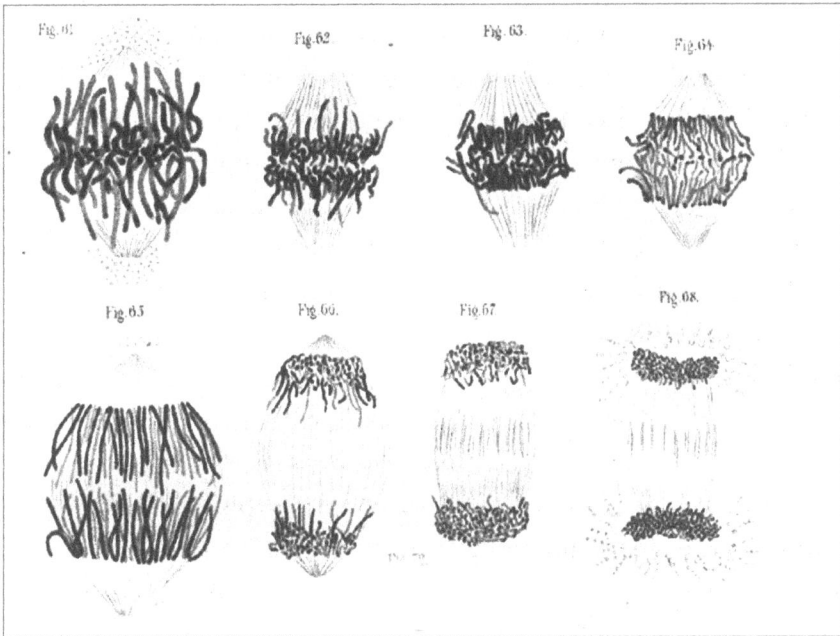

FIGURE 8. Drawings of different stages of mitosis, with chromosomes at the center of the spindle moving to opposite sides of the daughter cells. From Walther Fleming, "Cell Substance, Nucleus and Cell Division," 1882.

in exactly the right place for the type of cell that's dividing? How do they measure distances or count chromosomes and their strands so that half are transported to one side of the cell and half to the other? Is the placement of the spindle, where the chromosomes first assemble in mitosis, written in the genome?

There is much we still don't know about mitosis, particularly about the control and precise engineering of the spindle, centrioles, and chromosomes. But we can say that the answer to many of these questions cannot lie in the genes. Throughout the process of mitosis, genes remain inert, passive molecular structures. The chromosomes, on which genes reside, are copied and transported but have very limited involvement in what's happening. Instead, it's the organelles within the parent cell that sense space, balance forces, partition, and move chromosomes around. For example, we know that the timing of mitosis is not decided by one particular gene or even a collection of genes. Whether a cell divides or not relies on the assessment of many of its features: its volume, composition, stiffness, availability of nutrients, and signals from its neighbors. None of these can be directly linked to genes, though they will have an impact on the activity of the genome by determining the expression of certain genes whose products might be needed for a particular event—but the cell will decide what happens.

We are starting to learn how the cell does this; for the moment, however, we only have glimpses of the process. For example, a protein called mTOR plays a central role in gathering the information outlined above, determining when a cell will divide by controlling its growth. Found in all cells in the body, mTOR senses the cell's health and available nutrients and, based on this information, decides whether it should make more nutrients, divide, or die.[1] In a situation of stress, mTOR is able to read the state of the cell integrating many variables, measure them against the available nutrients, and, if possible, promote biosynthetic metabolism and growth. We still do not understand how it makes the decision.

In one dramatic instance, mTOR controls a process called *autophagy*, meaning *self-devouring*, in which, when stressed or running out of resources, a cell turns on itself and recycles its proteins and organelles, usually damaged ones, as a source of energy to stay alive. mTOR is a gatekeeper of autophagy. Usually it suppresses this process, favoring less dramatic ways to create mass and power, but if it assesses that there is not enough metabolic substrate for this, it will be switched off, allowing autophagy to proceed as an emergency system.

This is an example of a vital activity that happens far from the realm of genes. Of course, mTOR is encoded by a gene, but this gene only provides the instructions for assembling the protein. It is mTOR that does the heavy lifting that keeps the cell alive. The protein's function is determined not by the activity of the gene but, as with any sensor, by the protein's interactions with other proteins and the wider chemical environment.

The genome is neither an instruction manual nor a blueprint for making another cell. From the perspective of a cell, the genome is the catalogue of a hardware store with a vast array of tools, fixtures, and building materials from which cells can pick and choose. This analogy is appropriate because the cell calls upon genes in different ways as it needs them. Cells require both tools to perform particular tasks and materials to create stuff. A good example of a tool is the *transcription factor*, a protein used to transcribe genes. Transcription factors are used to activate the expression of genes, in the form of RNA, that have the code for other tools, including other transcription factors, elements of the signaling routes that cells use to exchange information, and the enzymes that mediate your metabolism or create the lipid building blocks of the cell's membrane systems. Some of the instructions are used to construct the cytoskeleton, centrioles, and also the channels that regulate the internal composition of the cell and transmit the electricity that powers the nervous system. Others will lead to the proteins that sit on the surface of cells and that cells use to stick to one another or survey their environment—the list goes on and on. The integration of all these structures creates the cell.

Against this array of activity and function, DNA is a collection of instructions waiting for a user, and this user is the cell. Don't get me wrong, this is important, but for all the complexity of the genome, genes are not actually running all that much of our daily lives. We are alive only so long as our cells are active—manufacturing hormones, distributing food and oxygen around the body, and conveying electricity, all of which are required to keep our hearts beating and our neurons firing. When these activities come to an end, life ends. Yet our DNA remains, and so long as it's protected from outside elements, it will continue to exist for thousands of years. The molecular stability of DNA is why scientists have been able to extract DNA from the remains of early humans living in Africa tens of thousands of years ago, compare their genomes with modern humans', and work out how people populated the world.

But they could not resurrect early *Homo sapiens*, or Neanderthals, from this DNA alone. While DNA carries a record of our history as a species, it cannot create a living individual.

For DNA to be meaningful, it needs a cell. Without a cell, the genome cannot express any of the information it holds. It takes cells to define what we are and how we are made. In fact, it takes two cells in particular.

A VERY SPECIAL KIND OF CELL

As we have seen from the organization and activity of neurons, not all cells are equal. Far from it. But among the two hundred different kinds of cells that make up the human body, one—the *gamete*—is exceptional. You will almost certainly know the gamete by the names of its two flavors: the egg and the sperm. The fusion of an egg and a sperm creates one, single first cell, known as the *zygote*, from which you, like all animals on earth, were built.

Gametes are ubiquitous in the animal kingdom, where new organisms are almost exclusively generated through sexual reproduction. At the heart of their job lies a trick of genetic accounting on the chromosomes, the packets of DNA in which our genome is organized. Most of the time, chromosomes exist in the nucleus as loose fibers, difficult to see even with a microscope. But when a cell is going to divide, chromosomes condense into thick structures that are visible and can be counted. As we have seen above, each organism has a specific number of chromosomes, but these do not seem to have any relationship with the organism's complexity.

As we have seen, chromosomes always come in pairs, which is why all organisms have an even number. And, as it happens, each chromosome comes from one parent. We have twenty-three pairs. At your life's inception, you received twenty-two chromosomes from the egg of your mother and twenty-two chromosomes from the sperm of your father. These twenty-two chromosomes were the same length. The last pair of chromosomes determined whether you were born with male or female sex characteristics. You got an X chromosome from your mother, who had two X chromosomes, and either an X chromosome or a Y chromosome from your father. If you got a Y, which is much shorter than an X, you were born male.

In each species, the number and pairing of chromosomes is important. People with three copies of chromosome 21 will have Down syndrome; those with three copies of chromosome 13 will have Patau syndrome, where the skull, brain, spinal cord, eyes, heart, or other organs do not develop correctly, typically leading to the baby's death in the first year after birth. We don't yet understand how a third copy of a chromosome leads to these syndromes, but it's clear that the number of genes matters.

The cellular art of accounting highlights what makes the sperm and the egg special. While all other cells in your body have forty-six chromosomes, a number that is replicated through mitosis, the gametes have twenty-three each, so that when they come together to form the zygote, the number forty-six is restored. This magic number (whatever it is for each species) cannot be tinkered with. Though the number of chromosomes differs, the same essential arithmetic happens in every animal, fungus, and plant on earth. This remarkable accounting exercise is produced by the process of *meiosis*.

Very early in development, a small set of cells, called *germ cells*, move to where the gonads will form. Like all the other cells at this stage, the germ cells have forty-six chromosomes. Through two divisions, they reduce that number to twenty-three. This is an amazing process not only because the cells count the chromosomes and their strands across two divisions, but also because, during this exercise, pairs of chromosomes exchange genetic material to ensure that you do not inherit exact copies of your parents' tools and building materials. I don't mean that we have different tools from our parents—we all have the same genes. I mean that the details of each of those tools will differ: a wrench will have a different handle, a saw, a different number of teeth, or a nail, a slightly different diameter.

At the end of the process, there are four cells, each with a full set of twenty-two chromosomes (plus an X or a Y), ready and waiting to build another human being.

If you were born female, the precursors of your eggs were generated by the fifth week of your development. They numbered about two million. Over the course of your childhood, many of those gametes died and were absorbed by the body, so that by the time you reached puberty, you had about four hundred thousand gametes left. At that point, only 10 percent were mature enough to meet sperm and become a zygote when, each month during ovulation, they were released one (or a few) at a time. The rest would continue to mature, or die, over the next thirty-five to forty years. By age fifty, when most women have gone through menopause, they have about one thousand gametes left. Typically, none of these will ever mature.

A mature female gamete, or egg, is a large cell, measuring 100 to 120 micrometers in diameter. That makes it as large or larger than the body of a motor neuron and about the width of a strand of hair, visible to the naked eye. It is a cell in all its glory, with a proper nucleus, mitochondria, and most cellular structures. The only parts missing are the centrioles, which are chucked out during meiosis because they're extraneous, and half of each pair of chromosomes. This supersized cell saves up stores of nourishment that must be held inside to fuel a healthy, fertile female gamete while it waits, potentially for decades, to be released.

The male gamete, or sperm, is the egg's opposite: a stripped-down version of a cell, with a densely packed ball of DNA and centrioles at one end and a large flagellum, or tail, at the other. The tail allows it to swim upstream through the uterus and fallopian tubes in search of an egg. Unlike females, males aren't born with gametes. Their gonads only start making sperm at puberty. They continue to create sperm throughout the rest of their lifetimes.

The fusion of human egg and sperm into a fully functioning cell is a race against time that takes place in the uterus. An egg will survive for only about twenty-four hours in the uterus, and a sperm can last no more than two days in the uterine environment without meeting an egg. But naturally, this situation doesn't involve just one lonely sperm and one lonely egg trying to meet. Thousands of sperm are swimming upstream seeking an egg, and those that are in the right place at the right time of the month have a chance of being the first to find it. The sperm will penetrate the tough protective membrane around the egg and inject its DNA and centrioles into it. And thus, the zygote, the first cell of the new organism, starts off in life with all the tools it might need.

The cooperation built into sexual reproduction is a curious fact of how some complex life-forms on earth have evolved. In all mammals, a new life requires that one of each type of gamete come together, because some genes in the male and female gametes are shut down in such a way that they need their opposite in order to fuse into a working, single cell. Some parts of the genome inherited from the mother are not active early in the development of an organism, but they're active in the genome inherited from the father, and vice versa. When scientists tried to breed mice from two mothers, they had to cut out large portions of one female gamete's chromosome, which would have been inactive in sperm, and then less than 15 percent of the resulting embryos—twenty-nine mice—survived to birth.[2] Trying to breed mice from two fathers was a failure: only 1 percent in the experiment survived to birth, most had developmental defects, and none made it to adulthood.

In building an animal or a plant, cooperation rules, from the day the first cell forms to the day the organism dies.

MERGERS AND ACQUISITIONS

It is sometimes said that how life originated is the big question in biology. Often when big scientific questions like this are raised, the person doing the asking has a particular idea of where the answer might be found. And over the course of the past century, theoretical and experimental energy has been devoted to working out how DNA and RNA molecules congealed in the primordial chemical soup and then found methods for copying themselves. Much of this research rests on the assumption that life is made up of active DNA and RNA molecules. Biochemist Nick Lane begs to differ somewhat and suggests that metabolism, the array of chemical reactions essential to power life, came first. From this perspective, nucleic acids are a coda to metabolism and how everything becomes encased in a protocell and encoded in a protogenome—details to worry about later. This is an interesting line of investigation, but in fact there are too many possible ways in which metabolism, nucleic acids, and membranes might have come together to kick-start living systems into being. Frankly, I don't believe we'll ever know exactly what happened. There are no fossils of those events, after all.

For me, life starts with a cell. And so, to my eye, the far more interesting moments in the story of life on earth occurred about 4 billion years ago, when the first cells came into being, and then around 2 billion years ago, when two classes of cells—the *prokaryotes* and the *eukaryotes*—diverged and began milling around the earth. These names come from the Greek word *karyon*, meaning *nut*, a way of referring to the nucleus. Prokaryotic cells are "prenucleus," or lacking a proper nucleus; eukaryotic cells have a "true," or well-formed, nucleus.

Relatively simple life-forms like bacteria are prokaryotic cells. They tend to live as individuals, feeding and fighting invaders and each other. Their existence is pure chemistry and movement. They often have a hard outer wall, which helps to keep them apart from other cells, even when they are living cheek by jowl in the same environment. Inside this shell, they have basic organelles, more like organized chemical factories—energy-creating and energy-transforming appliances, devices to search for food, motors to access it. Not an exciting existence, if you ask me. In almost all cases, their DNA consists of a few thousand genes wrapped up in a circular yarn that floats freely about the cell interior. They reproduce asexually, cleaving their case down the middle, untangling the two strands of DNA so that there's one in each half, and then using the single strand as a template to reform the original yarn.

The cells we have described above, the protagonists of this book, are eukaryotic cells, and, as we have seen, they look and behave very differently from the prokaryotic ones. These cells can exist as individuals, but we know them best when they form the social structures we call plants, fungi, and animals. They regularly interact with each other, eagerly adopting sexual reproduction and building large aggregates in the form of tissues, organs, and organisms.

The differentiation of eukaryotes from prokaryotes was originally explained as a slow process of evolution. Scientists assumed that some species of bacteria or other prokaryotic cells slowly accumulated mutations, thereby acquiring ever more complex systems of internal membranes, then a nucleus and other organelles, including mitochondria, which transformed them into eukaryotic cells. With two billion years between the appearance of the first prokaryote and the first eukaryote, there was more than enough time for evolution to happen in this way. However, when you study the structure of eukaryotic cells, a ridiculous, almost impossible fact stares you in the face: mitochondria look an awful lot like bacteria.

In 1967, a young, gifted biologist at Boston University named Lynn Margulis made note of this similarity and concluded that it wasn't a coincidence. In a seminal paper, "On the Origin of Mitosing Cells," she argued that eukaryotic cells are the product not of small, gradual changes but of a dramatic event, where one type of prokaryote swallowed another, and the two discovered they could work better together.[3] She called this process *primary endosymbiosis*—the first symbiotic relationship, where one organism lives inside another to their mutual benefit. Cannibalizations probably happened all the time; recall that prokaryotes spend much of their lives feeding and fighting. But in some proportion of cases, both the eater and the eaten prospered: the phagocytosing, or ingesting, cell provided shelter; the consumed cell provided an energy factory. She believed at least two other organelles in eukaryotic cells, including photosynthetic plastids, had originated through such an endosymbiotic merger.

Margulis submitted her paper to fourteen journals before it was accepted for publication by the *Journal of Theoretical Biology*, and when it came out, it was met with astonishment. Her critics couldn't believe her audacity in suggesting that Charles Darwin's laws didn't apply to the origin of much of life on earth. She could not be serious. Symbiosis in the natural world was incredibly rare! She was leaping to assumptions based on superficial resemblance! She and her ideas, deemed "unruly" and "exasperating," were swept aside. But her argument had a logic: the DNA in mitochondria is contained in a simple, circular yarn of chromosomes, just like the DNA in bacteria; mitochondria also have not one membrane but two. Perhaps the second membrane was a cell wall, the ghost of the ancestor bacterium's previously independent existence.

By 1978, the field of molecular biology had advanced enough that it was possible to prove that mitochondria have their origins in bacteria. They have their own patterns of growth and replication, distinct from the process of mitosis that reigns over the DNA in the cell nucleus. Now that DNA sequencing techniques are available, scientists have even been able to identify a division of bacteria that may have been the precursor of mitochondria, pseudomonadota, and to map genetic similarities to one subfamily, Rickettsiaceae. Modern Rickettsiaceae must live in a vesicle, or compartment, within a host cell.

This still leaves the mystery of the nucleus, the vault of the genome with its special membrane. Bacteria do not have a nucleus, so where did it come from? Perhaps after the endosymbiosis that created the mitochondria, the work of natural selection continued on its usual course, and the new cell, festooned with its bacterial partner, slowly collected all the accoutrements it has today. Another possibility, which has recently gained traction, has been put forward by David and Buzz Baum, cousins and biologists who work on different sides of the Atlantic. They believe that the win-win merger might not have been between two bacteria but between a bacterium and a

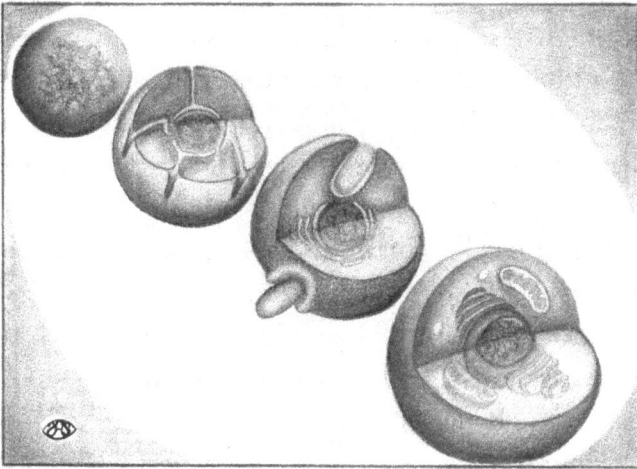

FIGURE 9. Eukaryotic cells emerged through the merger of two prokaryotes with different functionalities whose main features became organelles, most conspicuously the mitochondria. Further complexity, in the form of additional organelles, could have arisen by further mergers or gradual evolution of the internal structure.

member of a special class of single-cell organisms called *archaea*, prokaryotes that share some structural features with eukaryotic cells.[4] Like Lynn Margulis's work, the Baums' research took a lengthy trip on submission to various journals before it was published.

The Baums imagine that some ancient archaea may have had an external membrane covered in filaments that were used to engulf bacterial cells and bring them into their interior. On discovering that phagocytized bacteria complemented its own functional repertoire, the archaea retained them rather than destroying them. Over a longer sweep of time, the filaments that enveloped the bacteria were transformed into the precursors of the internal membranes surrounding eukaryotic cells' organelles, and the core of the archaea became the nucleus, with its own DNA tucked safely inside. As the Baums put it, this would mean that eukaryotic cells developed from "the inside out."

A growing body of evidence supports their view. One group of archaea—called the Asgard, after the home of the Norse gods, because their DNA was first sampled from Loki's Castle, a group of hydrothermal vents on the Arctic Ocean floor—contain genes only previously known to exist in eukaryotic cells. These genes are transcribed and translated into proteins that can be used to build the cytoskeleton, which helps to maintain the interior cell structure of the nucleus and supporting organelles. In other words, the inventory of the Asgard archaea's genomic hardware store overlaps with that of eukaryotic cells. Then, in 2014, Japanese researchers grew an Asgard archaean cell in the lab. This cell could only be cultured in the presence of other cells—another archaean and a bacterium. More excitingly, it displayed octopus-like protrusions similar to the filaments that the Baums imagined.

Endosymbiosis has emerged as the leading theory in explaining the origin of eukaryotes. But the example of the lab-grown Asgard archaean and Rickettsiaceae bacteria is not meant to suggest that eukaryotic cells emerged from one dramatic smorgasbord, and that was that. Instead, mergers between simple cells were almost certainly happening regularly, with different mergers playing out in the competitive environment. Margulis called this serial endosymbiosis. A hint that such mergers might not have been uncommon comes from the division between plants and animals. While mitochondria are the living fossil inside animal cells, plant cells appear to have arisen because cyanobacteria, which are capable of *photosynthesis*—the process that uses the light of the sun to create food and energy—were invited to the party in their ancestral eukaryotic cells. This tells us that a way for organisms to become complex and perform complex tasks is to combine different, let's call them skills, from different individuals in a cooperative manner.

These mergers are a sensible, and now we know correct, explanation for how a eukaryotic cell was put together. However, if one feature makes these cells what they are today, it is not simply their complex structural organization but rather what they do when the parts come together: the displays that we can see when eukaryotic cells move, stick to or separate from each other, come together to build tissue and organs, or divide. This is not easy to explain in terms of structure.

MORE THAN THE SUM OF THE PARTS

To understand how the parts of a cell come together to generate structures that can do more than one can glean from their individual makeup, we need to understand one important concept: *emergence*, a phenomenon whereby a combination of parts leads to activities not found in any of the individual parts.

If the best gene decoders were given a sequence of DNA from a previously unknown eukaryote and asked to guess what type of cell it was, they would have a hard time. Certainly they could not infer the cell's size or shape or the number and organization of its organelles. Nor could they describe how the cell moved or what it did. No surprise here, as all the cells of our body have the same DNA, even though there are many different kinds of cells. If our gene decoders were lucky and got some information about the RNA and protein content of the cell, this might give them a clue as to the tools and materials being used by the cell, which might tell them something—again, only if they were lucky—about what kind of a cell it was. This is because the form, function, and movement of a cell, particularly a eukaryotic cell, are created through interactions between proteins, RNA, and the lipids and sugars surrounding them. The structure of a cell is much more than just an assembly of its constituent parts.

The phenomenon whereby interactions between the component parts of a whole, a system of any kind—such as a flock of birds, a collection of water droplets, or the components of a cell—produces structures and behaviors that cannot be predicted from the assembly of its individual parts is called *emergence*. This notion is central to understanding what cells really are.

The first basic element of an emergent structure or process is that the components of the system *must* interact. By this I mean that when these elements come together and combine with each other, they exchange information about their states or structure. Furthermore, how this exchange proceeds depends on how they come together. As an analogy, there are several ways of putting three and three together. They can be added to give six; no big deal here. But if we multiply them, the outcome is nine, significantly more than their sum. They can also be divided, sort of canceling each other out to yield one. Such unpredictable outcomes result from the nature of the interaction between the two numbers. The second element of an emergent process is that the interactions follow very simple rules—addition, multiplication, or division in the case of how three and three might come together. Importantly, the outcome cannot be predicted merely from knowledge of the components: three and three could yield six or nine or one, depending on the interaction. This is why the outcome is said to "emerge," to "become apparent," as the dictionary says. Emergence underpins complexity.

Not all things around us result from a process of emergence, not even all complex things. In most places, three and three are six. Most machines in use today are complicated structures that require the precise assembly of many parts to function, but there is no emergence. Cars and airplanes, for example, are made up of thousands of components arranged following a blueprint that describes how each and every part should come together, defining the structure and behavior of the vehicle to ensure that the machinery works as intended. Think of the steering wheel of a car and how it controls the wheels, or the way the wing flaps on a plane work to lift it in the air or bring it down to the ground. The structure is no more than the ordered sum of its components. Any glitch is the result of something having gone wrong in the assembly, and one only has to go back to the blueprint, pinpoint the mistake, and fix it. This is not the case with emergent structures. Emergent structures do not have a blueprint.

Much of life is emergent. Its organization and behavior cannot be predicted from the way its parts come together; rather it is a consequence of their interactions. Take how a city grows and develops. A city planner, architect, or developer might start with an idea for how a district will be laid out. But then people start to live and interact there. A park develops a dirty trackway across it, as people take the shortest route—a so-called desire path—between two popular locations. This changes the character of the park. A new shop opens, selling unique goods at an attractive price. People who never used to visit the street now go there. Nearby shops appear and get more customers. Some of the people who work or shop at these businesses want to live nearby, and this leads to new housing developments. Desire paths, housing patterns, and hiring patterns change. Now multiply this by a million or five million individuals, each day of the year, across the landscape of a city, with terrain affecting where it's easy or hard for people to go. To "plan" a city is therefore to make simple rules and then watch how the districts evolve, making unpredictable changes in response.

Nature is emergent in the same sense as a city. The mock-gothic mounds built by termites in Australia, Africa, and South America aren't dictated by the queen. All of the thousands of termites are simply responding to the environment as they deposit their personal additions to their collective home. Yet these aren't just impressive, towering structures; the mounds feature a finely tuned network of pores that ensures hot air and carbon dioxide are carried out and cool air and oxygen are filtered in. Flocks of birds, swarms of bees, and schools of fish are also examples of emergence.

Behind all of these biological organizations and structures is the cell, which itself should be seen as an emergent structure. Nowhere in the genome will you find instructions for having exactly two centrioles or for the length and tension of the spindle that assembles before a cell divides. These features arise from multiple proteins coming together and clicking together to form structures, often as a result of interacting with other protein structures and membranes within the cell. These interactions multiply or divide the proteins' capacity to generate unsuspected outcomes. Whether a centriole gives rise to a cilium or a spindle depends on the proteins available at the time and their interactions, and some of the pertinent proteins are only activated in certain conditions or in the presence of other proteins.

The rules governing the interactions between proteins are quite simple—attach to this or repel that. It is the number of proteins in the cell, some forty-two million molecules in a simple yeast cell alone, that creates the bewildering, unpredictable complexity of interactions. Remember that there are only four bases (A, G, C, T) from which to provide messages through their organization in DNA. Structurally, not much can be done with the bases on their own, because they are almost identical in structure, and with their backbone of sugar and phosphate, they can only fashion a double helix. However, the twenty amino acids that form the basis of proteins are structurally very different from each other and can be arranged in 20^{20} ways—about 10^{26} (or one followed by twenty-six zeros). Further, the length and organization of a protein has no limit; this makes the possible architectures almost infinite. It is from the interactions between proteins, along with RNA, lipids (whose own molecular complexity is far too often forgotten), and (yes, it's true) *sometimes* DNA, that cellular structure and activity emerge. And it is from the interaction between cells that an organism emerges.

The mechanisms of emergence have been elusive and can easily be construed as magic or interpreted in the light of vitalism, a nineteenth-century notion suggesting that a mysterious, invisible force pervades life. However, emergence is a truly material concept. With the arrival of advanced computing techniques and machines, it's possible to take real measurements of proteins and their interactions and create, on a screen, a model of emergent processes like the division of a cell into two or the assembly of a spindle and its dynamics pulling chromosomes. For this to work, the interactions between proteins must be represented not by sums but by multiplications and divisions.

To some extent, the difficulty of understanding the emergent nature of the cell has been part of the reason the cell has taken backstage to the gene over the past century. The gene is well within our scientific comfort zone. We can describe its structure and rules of action in precise molecular terms. We can observe the effects of mutations in genes for eye or pea color and establish connections between genes and disease. From these we infer, sometimes successfully, the normal function of the protein encoded by particular genes. We understand genes and their workings so well that, recently, we have started to tinker with mutations in a very precise way and, in a small number of cases, to repair them and cure disease—for example, in the case of thalassemias, a group of diseases caused by mutations in the gene encoding beta globin that reduce the oxygen-carrying capacity of your blood. A cure has been devised using tricks from the CRISPR technology, a slick molecular tool, a sort of genetic scissors, that allows precise and controlled changes in the structure of DNA. Even so, most of the time, it is difficult to figure out what the function of the protein encoded by the normal gene is. Genes have information for components from the cilium to the cytoskeleton or the mitochondria, emergent structures that only make sense in the context of the cell. The cell is the unit of life, and emergence is its secret. At a higher level of organization, as cities emerge from interactions between humans and cells from interactions between molecules, organisms emerge from interactions between cells.

The intrinsic creative abilities of cells came together in numerous, unpredictable ways and gave rise to the organisms that colonized and still inhabit our planet. This jump, from single cells roaming ponds and riverbanks to multicellular organisms with division of labor and function and an ability to create and shape space, was not foretold in genetic code. The moment the first multicellular organism emerged, the balance of power changed, and cells found ways to use genes for purposes that single-celled organisms had not conceived of.

three

A SOCIETY OF CELLS

SOME THREE BILLION YEARS AGO, THE EARTH WAS A CAULdron of activity and experimentation. We cannot know exactly what was happening, but many scientists imagine a planet whose atmosphere had little oxygen. In such a world, prokaryotic cells would have feasted on hydrogen, nitrogen, sulfur, or carbon, creating oxygen as a waste product, just as many still do to this day. It took a billion years, but prokaryotes seemingly eventually generated so much waste oxygen that the atmosphere started to contain a significant percentage of this element. Though not nearly the 21 percent that exists in the atmosphere today, it made a difference, setting off a series of crucial transformations.

For certain cells, it was a boom time. Bacteria that were *aerobic*, meaning they converted oxygen into energy, were now living in a much friendlier world. Archaea, which by then might have swallowed up aerobic bacteria, were also finding ways to make the most of their new and changing environment while developing relationships with their guests. The change in the atmosphere had made it possible for life as we know it to emerge. A few million years passed, by which point a host of eukaryotic partnerships had developed, and things began to settle down. The alliances between

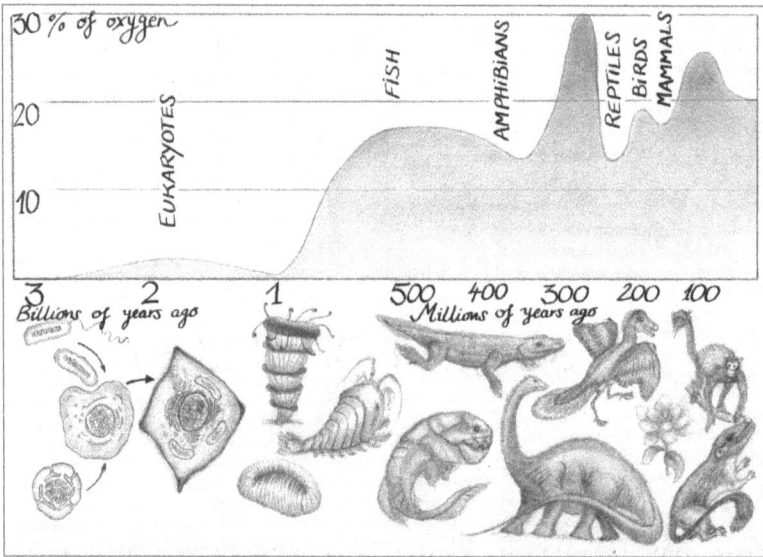

FIGURE 10. Timeline of events during the emergence and evolution of animal life on earth.

bacteria and archaea were just the beginning, allowing cells to reliably find food, create energy, escape predators, and, most importantly, reproduce. Many were living in anaerobic places—deep in oceans or muds—where they found shelter from oxygen, which was a danger to cells that hadn't found a way to use or expel the gas. But some harnessed oxygen to fuel their existence. While the majority of life on earth was still made up of prokaryotes, the world of two billion years ago was also populated by a diverse population of strange single-celled organisms that today we call *protists*.

We can get a glimpse of this ancient, ancestral world by following Antonie van Leeuwenhoek's example and putting a sample of water from a pond under a microscope. Each drop is teeming with life. Cells with the appearance of fantastical monsters swim in and out of our field of view, their outer membranes flocked with tentacles, pincers, and cavernous, mouthlike openings that they use to hunt for food. Other cells anchor themselves to nearby surfaces with a thin stalk projecting from their main body, creating the appearance of a minuscule mushroom inhabiting a microbial forest floor. Such stalks feature large pores that take in water and filter it for food. There are cells wriggling around like snaky tubes, cells that appear to use vacuum cleaners to hoover up their lunch, cells with shoelike appendages that tiptoe and tap-dance around as they explore their environment.

Most of these creatures are translucent, though some have a green tinge. This color is indicative of one of the many intriguing partnerships between archaea and bacteria, giving some cells the ability to turn sunlight and water into energy. Ancient eukaryotes like this were the ancestors of plants. Those unable to create energy from light and soil needed to eat other animals and plants for nourishment.

Eukaryotic cells continued remaking the earth, increasing the oxygen in the air and thus the tools needed to survive on or near the planet's surface as new experiments in cooperation took shape. Some cells learned how to form colonies, which could offer protection against predators and starvation. As individuals they were susceptible to being slurped up by attackers and had to constantly vie for sustenance, which diffused away if they did not consume it immediately. By joining a colony, cells not only made themselves a more formidable challenge to attackers but could corral and hoard nutrients for use in the future. The ability to come together in "herds" allowed cells to cooperate in other novel ways, as in the case of the genus *Volvox*. These green alga cells live in colonies where individuals gather into a sphere and dedicate themselves to one of two roles: reproduction or movement. Those on the surface of the sphere have cilia, which they beat in sync to move together as a whole, and photoreceptors, which they use to navigate toward light. A smaller number of cells in the interior of the colony have the capacity to reproduce either sexually or asexually. It was an extraordinary turning point in the history of life on earth for individual cells to augment some functions and shed others, opting to depend on neighboring cells to step in to make up for their deficits in order to survive and thrive together.

We have to suspect that some early experiments in multicellularity were short-lived, a mutually beneficial arrangement that lasted for a few moments, maybe days, but didn't become permanently and necessarily interdependent. Bacteria, for instance, have been observed cooperating in large communities, called *biofilms*, in order to survive harsh environments, like the thermal vent of a deep-sea volcano or the gut of an animal. But each individual bacterium still maintains the freedom to move and feed and reproduce as it pleases. It can exist on its own.

That's not the case for those special organisms that turned multicellular colonies into a perpetual way of life. In time, some cells living together reached the point where their progeny remained together, attached to each other after division, and successive generations ceased seeking their own way in the world. Why this happened, we don't know. There may have been some mutation in a gene with the code for an existing protein or the invention of a new gene that allowed cells to remain together after division. In evolution, something that works and is to the advantage of the individual will survive. The glitch, whatever it may have been, was copied to the parent cell's daughters and granddaughters. In time such glitches became more complicated—some descendants used the tools for movement, while others used the tools for reproduction—creating complexity.

By the time oxygen was in abundance, cells living in a mass were becoming increasingly adept at carrying out different functions, which helped the aggregate survive—some with even more specialization than those of the *Volvox* alga. Some cells specialized in gathering oxygen, others in foraging for food, still others in providing protection, the ability to move, or reproduction. Individual cells could no longer survive without the support of others, and multicellular organisms had emerged at last.

Cells' newly found ability to work in groups that emerged from a single cell is called *clonal multicellularity*. Confederated in this way, descendants of a single original cell develop the ability to cease acting like identical clones and to specialize, differentiating themselves from one another. In the millions of years that followed the advent of clonal multicellularity, cells learned to use the genes they had to interact with each other, to read space and use those interactions to create shapes. It is no exaggeration to call this a learning process, because a true learning process was going on. Complexes of proteins were able to read the environment of a cell and react to it, and this complex was passed on to successive generations. New mutations expanded the repertoire of tools available in the genome, sometimes with the emergence of new genes coding fixtures and gadgets that would be used to shape cell aggregates in space and time, which became linked to different tools.

At this moment the relationship between genes and cells started to change as the conditions were finally in place for the ancestors of plants, fungi, and animals to be born.

WHAT IS AN ANIMAL?

The differences between plants and animals may appear obvious to the eye, but such classifications are never as simple as they seem. Aristotle mused extensively about animals' organization and function in hopes of devising a system for defining them as separate from plants. For Aristotle, a distinctive feature of animals is their ability to actively move themselves from one spot on earth to another, which contrasts with the passive movements of plants. Animals, he saw, also have a capacity to change, that is, to transform themselves, particularly during development, when a mass of cells is progressively shaped into a form in keeping with an individual species. Of course, Aristotle never used the term *cell* or *multicellular* in his writing, largely because the word *cell*, not to mention the notion, did not yet exist. Nonetheless he observed that an animal emerges from a formless mass of matter in unpredictable ways. Something strange was clearly at work in the makeup of animals—and the argument over their nature was just beginning.

For centuries after Aristotle, the line between animals and plants defied easy definition, and every time scientists and philosophers seemed to settle matters, an exception emerged. Take the discovery of hydra, a little tube of cells about thirty millimeters long with two well-defined ends: one with many tentacles and a mouth that serves as a feeding organ and, at the other end, a "foot" that is used as an anchor to the ground. The Swiss naturalist Abraham Trembley noticed it in a pond while walking by the estate of Sorghvliet near The Hague in 1741. Puzzled by its appearance and uncertain if it was a plant or an animal, Trembley decided to seek a solution by cutting the organism in half. After all, plants were known to regenerate, while animals did not. In this respect, it acted like a plant, growing into two versions of itself from the severed halves. At the same time, the little thing moved and fed like an animal, and, surprisingly, when repeatedly cut in pieces, it always re-formed itself; it seemed to be unkillable.

Hydra is an animal, but its differences from plant life are subtle. Though both plants and animals emerge from a zygote, the union of male and female gametes, and then go through multiple cell divisions in which individual cells within the organism take on different structures and functions, two additional features distinguish plants and animals and also apply to hydra: construction and sustenance. Plant cells are like bricks: as they multiply, they are added in a geometric fashion to preexisting ones, sometimes at angles that create variety in form but without much change in the relative position of individual cells; plant cells don't move. In contrast, animal cells come in many flavors, are pliable, move, and combine with each other in a large variety of ways. Another, crucial difference relates to the fact that plants manufacture their food and energy from light and water, while animals must get their energy by consuming other organisms.

But even these two distinctions run into exceptions and special cases. Where do protists fit into this scheme? Are they more like animals or plants? To scientists' relief, even the smallest and strangest organism has a specific DNA sequence. The strings of nucleic acids—the As, Gs, Cs, and Ts—are now simple to read and can be measured. Surprisingly, these measurements are sort of a barcode that can reveal the degree of relatedness between species and animals' deep ancestry. This can be done by looking at genomes, side by side, and comparing the sequences and positions of the letters lined up in a chromosome.

The degree of sequence similarity between individuals of the same species reveals how close they are genetically. Your genome differs from your mother's and your father's and that of any siblings you have by around 50 percent—though most likely more or less. Notwithstanding this, all human genomes are very similar to each other. So the differences we can see between individual people only pertain to a few million of the 6.4 billion letters in the genome—enough to create individual barcodes. A look at these differences can identify your relatives, close and distant. This is indeed how companies like Ancestry, MyHeritage, and 23andMe offer individuals who have taken a DNA test access to a list of people who may be their second to third or fourth to sixth cousins based on the presence of the same string of letters on some chromosomes. This technique also helps solve crimes: DNA found at a crime scene can be entered into a database to locate relatives and narrow down the identity of the culprit. In much the same way, we make comparisons across and within other species and build a "family tree" of life on earth.

The basis of the technique is simple and highlights the barcode nature of our DNA sequences. Take the nine-letter sequences AGGCTATTA and TCGCTATTA. These are very similar, differing only in the first two letters. This suggests a degree of closeness between two individuals who each carry one of these sequences in the same area of the same chromosome. Compare these to a sequence like CTGCTGAAT, where six of the letters are different from both AGGCTATTA and TCGCTATTA. The bearer of this third sequence could plausibly be related to the other two, but more distantly, because there's been more mixing of genetic material across more generations of reproduction. Using computer algorithms that start from the assumption that the differences reflect changes from an original, common sequence, scientists have compared the genomes of many species. This has yielded a family tree of all life that is both more and less complicated than previously imagined.

This became clear when scientists started to reckon with the question of where archaea fit on the diagram of life. Based on Carl Woese and George Fox's 1977 analysis of archaea's RNA coding genes, scientists realized that archaea were so different from bacteria that they should have their own kingdom. There were some commonalities even so: the same molecular evidence helped to establish Lynn Margulis's theory of endosymbiosis by showing that mitochondrial DNA is related to bacteria, and nuclear DNA has similarities with archaea's.

Further genetic studies have found that the eukaryotic neighborhood is made up of over 8 million species. Only a few have been identified so far, including about 200,000 species of protists, 400,000 of plants, up to 1 million of fungi, and an estimated 1.5 million of animals. These groups are defined by similarities in their DNA, and the DNA of those organisms we call animals is more closely related to that of other animals, and to us, than it is to that of plants, fungi, or protists. This furnishes a sense of comfortable familiarity, but it does not tell us why each kingdom's organisms differ in how they organize the functions of life. To find clues, we must look more closely, not just at their genetic code but at the tools that genetic glitches created and allowed them to inherit. The emergence of these new tools was a crucial feature in the emergence of multicellular organisms.

ORIGINS: NEW TOOLS

To understand why, rather than how, organisms differ from each other, we need to look beyond the strings of chemical letters and compare the catalogues of genes, the actual words and sentences that can be found in DNA. This has allowed us to learn that the closest known nonanimal relatives of animals are a group of aquatic, single-celled organisms, lumped into the jumble that we call *protists*, featuring a distinct shape and organization. Like the mushroom-shaped cell found in pond water, these creatures have an ovoid body resembling a cap. Instead of a stalk anchored to a surface, they have a flagellum, or tail, extending from the cell body. The cytoskeletal proteins surrounding the flagellum form a funnel, giving the impression of a collar. Because of this, these distant cousins of animals are called *choanoflagellates*, from the Greek word *choano*, meaning *funnel*.

Choanoflagellates' DNA sequence appears to place them closer to animals than to any other protists, though it is the repertoire of genes at their disposal that clinches this relationship. Choanoflagellate cells exhibit many of the same proteins that make up animal cells, including proteins used to stick to other cells to form a colony, should this be advantageous. Making use of these tools, choanoflagellates have been observed organizing themselves into palisades where they cooperate to gather food. They also have proteins that can change their body's shape by controlling genes that code for elements of the cytoskeleton and surrounding themselves with an *extracellular matrix*, a protective environment composed of enzymes and gluey materials secreted by the cells themselves. The extracellular matrix is a very important component in multicellular organisms, not only providing mechanical support for the cells to organize their assembly into tissues and organs but also containing signals that cells use to read and shape the space around them. Some choanoflagellate species even have the capacity for cells to focus on particular functions during different stages of the life cycle. What separates them from animals is the genes they lack. In particular, they seem to be missing genes coding for elements of the signaling networks that mediate intercellular communication and also large swathes of transcription factors associated with creating the two hundred or so different cell types that are typical of our animal collective.

In the melting-pot community of protists, choanoflagellates have close cousins that also share some features with animal cells. For example, *Capsaspora owczarzaki*, the only species in the genus *Capsaspora*, also has access to the collection of tools and building materials that characterize animal cells, including elements of the *epigenetic machinery*, that system of chemical modifications in DNA and proteins that affects gene expression and is missing in prokaryotes and choanoflagellates. Among the tools in *Capsaspora*'s genomic hardware store is a copy of the gene used to make the protein Brachyury, which Nadine Dobrovolskaya-Zavadskaya discovered was related to the short tails and spines of mice. This is very surprising: What can this organism have in common with a mouse? When *Capsaspora*'s gene for *Brachyury* is placed in frog embryos, the frog cells are able to use the tool as their *Brachyury* gene, even though it's from a single-celled organism on a distant branch of the family tree of life.[1] Even developmental biologists find this surprising: *Capsaspora* does not have a spine, of course, or a tail of any sort. It doesn't use the protein product of the gene for these purposes. Remember, a cell uses the protein as it needs, based on its cell type and species. As we shall see, we now know that Brachyury allows cells to move, and perhaps this is its purpose in this ancestor of ours.

From the other side of the divide, the nearest animal relation to protists appears to be sponges, though at first glance they do not look much like animals. Carolus Linnaeus classified them as plants, in the same area of his plant kingdom as algae, and for a long time this classification was accepted as a matter of common sense. After all, sponges settle themselves on the seafloor and grow in branches, behavior that superficially resembles a plant's. But even during Linnaeus's lifetime, zoologists began to argue that sponges were animals based on how their bodies moved. And when you take a closer look at a sponge, you can see that they fulfill all the conditions for being an animal: first, they reproduce, either sexually or asexually, but in both cases starting with a zygote; second, to create energy, they eat microbes; and third, they have at least three different kinds of cells, each with a functional specialization, that result from interactions between the cells and unfold as the organism grows and matures.

Sponges forage for their food by generating water currents through the coordinated movement of special flagellated cells called *choanocytes*, or "collar cells." Choanocytes have a structure suspiciously similar to that of choanoflagellates—a caplike cell body containing the nucleus and a collar from which protrudes the tail.

So what makes sponges different from a colony of protists? Most importantly, they exhibit a clonal cellular diversity: their different cell types derive from a single cell and remain attached to each other. But there are also hints in the proteins encoded in their DNA: they have information for most of the tools and fixtures that are missing in choanoflagellates but present in animals. In particular, sponges have access to the signaling kit that allows animal cells to talk with each other, including the proteins behind the acronyms and nicknames BMP, Notch, Nodal, Wnt, and STAT. These proteins simply don't exist in choanoflagellates or their relatives, and their main role is to mediate communication between cells. We shall see more about them later, but as mutant screens in different organisms have shown, cells must have access to these proteins in order to build an animal body. In their absence, development fails. They are a unique feature of animals.

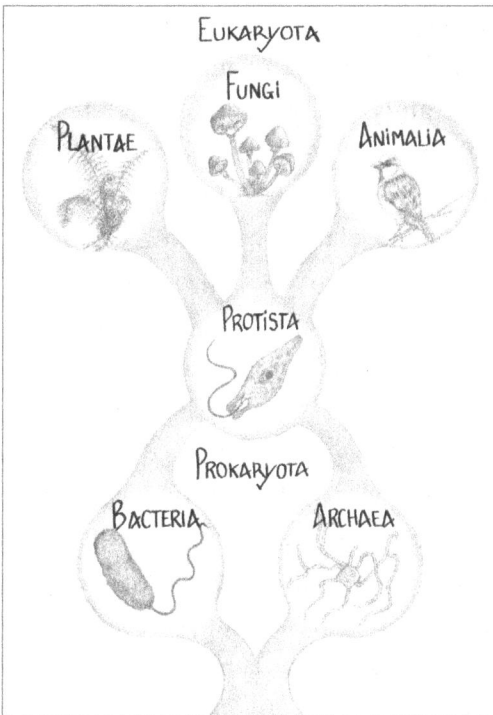

FIGURE 11. "Family tree" of life on earth based on genetic analysis. The prokaryotic bacteria and archaea appear to have combined to create eukaryotic protists, while two types of prokaryotic bacteria appear to have combined to create the ancestors of plants. Both fungi and animals have protist ancestors.

Scientists have found fossilized sponges in deposits dating to six hundred million years ago, a time when there are no hints of any other animals. These fossils appear very similar to sponges living today, so we can assume that sponges' genome has remained about the same over this long time horizon. This reveals their place at the very base of the animal family tree.

Between living choanoflagellates and sponges, scientists have found no closer genetic cousins—at least so far. It's difficult to imagine such a big leap in how cells organize themselves. The fossil record supports such a gap, but the fossils may not offer a perfect reflection of life on earth so many million years ago. Fossils of early choanoflagellates and sponges might lie packed deep in layers of sediment that were once the floor of a shallow sea. Or these ancestor organisms may have existed for such a brief time or in such a hostile environment that they left no fossil trace. We know from the clues to their ancestry tucked into their DNA that choanoflagellates must have existed before sponges, but researchers have not yet found any fossilized choanoflagellates.

This does not mean that the appearance of the first animal necessarily came all at once through some sort of multiple mutation miracle; there were almost certainly many intermediate steps in the form of ancestors to today's choanoflagellates and sponges. The emergence of animals was likely the result of a process of experimentation, perhaps of organisms associating with each other like bacteria and Archaea at the birth of eukaryotes. Regardless of how it happened, it certainly involved cells hoarding genes, inventing new ones, and trying out different ways to use the proteins and RNA encoded in the genome to build more protective and energy-efficient communities with other cells. The emergence of animals clearly came during a period of creativity at the genetic level, where new tools, most especially transcription factors and signals, were being generated through mutations and combinations and then trialed by cells. Perhaps, in the same way that eukaryotic cells were created through a merger of two or more organisms, multicellularity arose through a combination of several unicellular ones.

For now, we know only that the experiments in multicellular cooperation that led to the animal kingdom bubbled up between two billion years ago, when the first eukaryotic cells emerged, and six hundred million years ago, when sponges first show up in fossil deposits. Although these mind-numbing numbers represent incomprehensible spans of time, when you think about it, billions of years is quite long enough for such unlikely leaps to transpire. But then, very shortly after this moment of divergence, we start to see organisms in fossil deposits that look like certain types of marine animals living today—the beautiful, luminous ctenophores, or comb jellies, and the cnidaria, including jellyfish, sea anemones, and hydra, that so surprised the naturalists of the eighteenth century. With cnidaria comes the debut of *Hox* genes that some hail as the real hallmark of animal life.[2]

Whenever it happened, however it happened, soon after sponges appeared on earth, eukaryotic cells set about exploring ways to use the new genes at their disposal in groups to create and sustain life in a way that had not been conceived before. Against a backdrop of developing plants and fungi, animals prospered. Working in parallel and in partnership, cells began to change the face of the planet as they had done before with the atmosphere.

DELINEATING TIME, CONQUERING SPACE

Imagine all those protists in the lagoons and soils of a million years ago. Small, single-celled creatures milling about looking for food and chasing each other, surrounded by huge amounts of empty space. The only thing breaking up their monotonous existence would have been the occasional change to the environment and the opportunity to come together with some similar cells to form a multicellular aggregate, almost always as a ball of cells, where they might fall into a division of labor based on their position, à la *Volvox*. Still, these collaborations had sharp limitations. If the colony of protists grew, it just became a bigger ball, never changing shape or form.

This is very different from the life of an animal. To be sure, animals also start as balls of cells, perhaps in an echo of ancient protist aggregates, but in contrast to their cousins, animal cells mold increasing cellular masses into different shapes that perform different functions: tubes that absorb food, branches that exchange gases, chambered vessels that pump blood. This happens for two reasons: first, animals have the ability to generate a range of different cells from a single originating cell, and second, these different cells can do much more than just stick to their neighbors and form a ball like a protist—they can organize themselves into tubes, sheets, and fibers, loosen up, and move to a different location where they can start a new collective linked to the original one. Their ability to create a wide range of structures explains how they build a heart, digestive tract, and gills or lungs, as well as more unique and awe-inspiring attributes, like the neck of a giraffe, the wings of a condor, or the tusks of an elephant.

I believe that, from an architectural perspective, animals' particular approach to multicellularity is achieved through cells commanding time and space. Let us deal first with time.

For the billions of years that preceded the appearance of eukaryotes on earth, time—biological time—did not matter much; maybe it even did not exist. If time is perceived as a sequence of events, bacteria experience time but without a direction. Their lives are open-ended. They move about, divide, and sometimes form spores that freeze their existence, but for them, time lacks a clear direction. The appearance of protists—we assume they emerged before fungi, plants, and animals—changed things. Most protists have life cycles, and this brings in the sequences of events that are necessary to create time with past, present, and future. This becomes more apparent and intense in animals and plants, as from the moment that the zygote starts dividing, through the diversification of the cells spawned in this process, to the emergence of the organism, events follow irreversible sequences, each with a beginning and an end, all glued together by the invisible arrow of time. This is not the time of clocks that tick as the earth spins on its axis and travels around the sun. It is a different kind of time associated with the activities of cells. The two times, the astronomical and the cellular, however, are linked by what is called the circadian clock, which adapts cellular time to astronomical time to make our lives what they are. At the core of cellular time is the chemistry of the activity of proteins, which, deep within the cell, do the work of transcribing DNA into RNA.

Transcription factors, the main tools of the cell, bind to specific sections of DNA that control which portions of DNA are activated and, after the process of translation, which proteins become available to the cell "downstream" in time. For example, imagine that a cell in a developing animal needs to have a flagellum in order to fulfill its functions, as human sperm do. Let's say gene A is activated when the cell needs to make a flagellum, transcribing its code into RNA, which gets translated into a protein that then somehow leads to the expression of a subsequent set of genes B, C, D, and E; these, along with triggering one another like a set of dominos, also create the components that will be assembled into the flagellum. Each event in this sequence of molecular events has a duration associated with the chemical reactions required for it to happen. The sequence from the activation of gene A to the assembly of the flagellum therefore creates directional time. This happens in many biological processes. As a result, the schedules associated with temporal patterns of gene expression are called *genetic programs*, because the order in which genes can be expressed is regulated, or programmed.

Genetic programs did not begin with animal cells. Indeed, all prokaryotes and even viruses have genetic programs. But animals stand apart for their genetic programs because most of them—the ones we are interested in here are associated with the emergence of different cells—run sequentially, one after another, and in parallel in the different cells that make up the organism. As I have said, the core of these programs is the sequence of events created by the chemistry inside the cell, but then something needs to coordinate these across cell populations and, moreover, across different cell populations within the organism. We do not yet know how this happens, but we do know that the cell is responsible for it and that the signaling kit that we can see for the first time in sponges has much to do with it. Once again, genes and their chemical actions are under the control of cells.

The genetic programs that fuel the shaping of multicellular organisms are likely to be an elaboration of certain genetic programs found in protists, which change in appearance over the course of their single-celled lives. How this might happen can be gauged from one such protist, the humble slime mold. These creatures look like plants or fungi, but they are protists, happy to forage as single cells so long as they have food around. Things get interesting, however, if resources become scarce. In response to a lack of food, they signal to each other using a chemical called cAMP, a primitive version of the more sophisticated signals that will come later, which instructs them to come together into a large aggregate. After assembling together, the cells differentiate into one of three forms: foot cells, stalk cells, or spores encapsulated into what is called a *fruiting body*. In the resulting structure, foot cells find an anchoring spot, and stalk cells lift the spore cells up so that they can be carried away, on the wind or by a passing animal, to germinate somewhere that food is more available. Once the spores have been released, the foot and stalk cells die. This strategy of generating different types of cells, with some dying while others continue living, was passed on to the first animal cells.

While genetic programs generate what is needed to create different types of cells and a sense of time by dictating a strict chain of events, they are not enough to beget an animal on their own. For this to happen, the different cells have to come together into a functional structure, an architecture that conquers the space around them. Furthermore, the timing of those programs, which differ for the different kinds of cells, has to be coordinated to create a functional organism. The generation and conquest of space is a unique achievement of multicellularity.

When cells multiply, shape themselves together, and move relative to each other, they create two kinds of locations in space. One

FIGURE 12. The life cycle of a slime mold. Starting on the right and continuing clockwise, slime mold cells live as single cells. When the environment requires it, they form an aggregate. The aggregate elongates into a sluglike formation, then migrates to a better location. Once it's found a hospitable place, foot cells anchor to the ground, and stalk cells grow upward to form a fruiting body. The spores are dispersed, the foot and stalk cells die, and the cycle starts again.

is a place for themselves in the environment, while the other is a sort of "inner space" between and among the cells, over which they hold dominion. While protists can clearly move and can do so in sync, as slime mold aggregates do when they form a fruiting body, they only ever do this on a temporary basis. Animals, on the other hand, command space, establishing long-lasting "inner spaces" that they can continue to shape.

In this way animals operate much as developers do when they find an empty plot of land and put a structure up on it, and their work has similar consequences for the world around them. When a developer erects a house or an apartment block, an office build-ing or a railway station, the structure not only transforms the plot on which it sits but also has consequences for neighboring land and can potentially have cascading effects on how the whole community is organized. New traffic patterns emerge in response to the new building. New amenities must be put in place to serve those who live and work there. Towns become cities and transform a little bit more with each new development. In this way, animal cells have used their ability to organize themselves to transform both their outer and inner spaces, evolving from the simple structures of sponges and jellyfish that first appeared about six hundred million years ago into complex, long-term confederacies.

The most exciting experiments in multicellular engineering occurred around 540 million years ago, during a 20-million-year period called the Cambrian Explosion. At this time, an abundance of new animal life-forms emerged, only a few of which escaped extinction. We know this from fossils found in the Burgess shale, in the Canadian Rockies of British Columbia. In *Wonderful Life* (1989), biologist and evolution champion Stephen Jay Gould described the richness of life found in the Burgess deposits with his characteristic enthusiasm. "Consider the magnitude of this," he wrote. "Taxono-

FIGURE 13. Panorama of life in the Burgess shale about five hundred million years ago.

mists have described almost a million species of arthropods [which includes insects], and all fit into four major groups; one quarry in British Columbia, representing the first explosion of multicellular life, reveals more than twenty additional arthropod designs!"

We cannot extract DNA from fossils, but the appearance of these structures can tell us something about the nature of animals during this period. In the Burgess shale we see the forms of sponges and related organisms with strange designs—forerunners of corals, crayfish, and insects and massive, earth-burrowing, wormlike animals—in which amorphous shapes have been transformed into highly symmetric structures. Some animals have familiar body parts: antennas, legs, fins, shells. Others seem to be lifted from "the set of a science-fiction film," as Gould observed. No wonder one genus of mollusk from the shale was christened *Hallucigenia*.

On the other side of the Cambrian Explosion, one design emerged as a major winner, joining sponges and jellyfish (and, of course, plants) in their conquest of the planet. This was a segmented structure, with a feeding organ at one end and an excretory organ at the other. Unlike sponges and jellyfish, which structured themselves according to radial symmetry, this structure exhibited *bilateral symmetry*, or mirrored left and right sides. This is why animals with this structure are referred to by the name *bilateria*. Each segment in this design could change independently of the others. For example, limbs could take the form of legs, arms, fins, claws, or wings, and an animal could have any number of them, as long as bilateral symmetry was part of the design. Like most animals on earth, we humans are bilaterian. More than 90 percent of identified animal species are bilateral, from the 80 percent that are worms, insects, and crustaceans to the 4 percent that are mammals.

Experimentation continued apace. The resulting creations were so successful that they conquered water, land, and air. What fueled this creative exploration of and experimentation with form and function? The standard answer in the textbooks is that changes in genes instigated new shapes and forms in animal life. And there is little doubt that new genes play a role in animal building, with the

Hox genes, which provide signposts for the order of body segments in every animal, as a prime example.

Still, questions remain about the relationship between genetics and multicellular structures amid this profusion of forms. After all, if genes alone are responsible for new designs, why do we see the same genes across species that feature wildly different aspects and organizations? We should not expect that, when planted in a fly, a human *PAX6* gene will create fly eyes rather than human eyes. Genes likewise cannot explain why identical twins, who have the same DNA, aren't exactly identical or why my right arm is ever so slightly longer than my left. Nor can genes fully account for when cells in our body stop growing. They don't even distinguish left from right—a crucial omission, given the bilaterian explosion underway. The chemical information enshrined in DNA is also silent on the numbers and proportions of cells that should dedicate themselves to specific organs and tissue types. Instead, cells must sense each other and the space around them, monitoring how many of them are co-existing and where they are within their part of the whole.

One important feature of genes, the *transcriptional control region*, is pushed to the front as an explanation for the emergence of novelties in the evolution of animal forms and the different uses of genes. These bits of DNA determine when and where a gene should be expressed; they are the handles the cell uses to decide what to do with the gene. These are regions of DNA that do not code for RNA and are landing pads for transcription factors. It is thought that changes in these handles are an important way to redeploy genes for new functions. As much that we know about genes, this is not in doubt, but it is proteins, not genes, that give cells the ability to organize themselves in time and space in new ways. The emergent structure and function of cells controls the subsequent use of genes, not the other way around, and control regions are the means for the cell to access genes. How else can we explain the fact that many genes from other organisms remain so effective after being imported into a different species?

Understanding how animal (and plant and fungal) life emerged demands that *we see genes not as the instructions or blueprint for an organism but rather as the instructions or blueprints for the tools and materials that cells use to build organisms.*

With the dawning of the animal kingdom, genomes grew larger, giving cells more genes to work with. This opened up enormous possibilities but didn't actually make those possibilities into reality on its own. After all, the DNA was locked up tight in the nuclei of cells. In contrast, cells interact with their world. They live in four dimensions (space plus time), pushing and pulling themselves through space—albeit so slowly that you, as an organism, cannot feel these microscopic forces. They can mold themselves into a variety of shapes and functions, and therein lies their power.

Still, to use the products of those genes properly, cells needed another set of tools to sense and decide how best to deploy them. So it was that, to truly unleash their creative power, cells made use of—or, in the way *evolution* does, probably invented or discovered—signaling systems: BMP, Notch, Nodal, and Wnt, which arise at the same time as multicellular organisms.[3] How and whence these molecular devices came into being are not yet known, but their existence allowed cells to play with and master space and time.

Just as cells are emergent structures, the unpredictable result of interactions between their component elements, multicellular organisms too are emergent structures, the unpredictable result of interactions between their component cells and of interactions between those cells and their environment. Command over time and space is only the beginning—armed with signaling systems, cells are able to monitor and control the spaces they create in response to changing conditions that they sometimes create themselves.

The arrival of plants, fungi, and animals redefined the role of genes in the organization of life on earth and, in particular, their relationship with cells. With the advent of signaling systems, evolution gifted cells with the tools to remake their world through exchanges of information and, importantly, with the ability to control the activities of genes in space and determine the pace of the schedules that their programs generate: cells use and control genes. To understand how this happens and its implications, we need to revisit one of the most popular and pervasive notions in biology and the bedrock of the field's gene-centric view.

THE SELFISH GENE

"What can be more curious than that the hand of a man, formed for grasping, that of a mole for digging, the leg of the horse, the paddle of the porpoise, and the wing of the bat, should all be constructed on the same pattern, and should include the same bones, in the same relative positions?" Charles Darwin wrote these lines in *On the Origin of Species*, published in 1859. Darwin marveled at the variety of form and function across species, which all seemed to have such similar building blocks, and proposed that this resulted from a process not very different from that used by people selecting traits to generate breeds of, say, sheep with superior wool or dogs with a nose for hunting. However, Darwin suggested that rather than a breeder picking and choosing traits that become strengthened in each generation, nature provides the filters in the form of the environment. Starting from a simple structure with an intrinsic potential to adapt and vary, a shape emerges—hands, feet, legs, paddles, or wings— that, if best adapted to the requirements of the surrounding environment, will be propagated.

Darwin called this process *natural selection*. It involves a progressive adaptation to the environment through the stepwise modification of certain characteristics from one generation to another. It works relentlessly, often imperceptibly, across the timescale of our own personal lifetime. But over many millions of years, it effects enormous transformations and has allowed multicellular life to gradually move from the oceans onto the land and into the sky, opening up new niches for organisms to find food and reproduce.

Yet the idea of natural selection led Darwin into a new predicament. He understood what happened, but he still could not fathom a material source for the variations he observed. Although Darwin and Gregor Mendel were contemporaries, Darwin did not know about genes, and I am not sure that this would have helped him. Not until the turn of the twentieth century, long after Darwin's death, would a cohort of scientists recognize that the source of variation in nature lies hidden in the genes. After decades of further scientific elaboration and popularization, genes became accepted as the defining ingredient in our makeup, the determinant of development and identity, and the encapsulation of who we are as individuals that they are today.

One person has done more than any other to put the notion of the gene at the center of our understanding of life on earth and our being: Richard Dawkins. In 1976 Dawkins published *The Selfish Gene*. The title captured people's imagination, and the slim volume became one of the most popular books on evolution. In a 2017 poll conducted by the Royal Society, *The Selfish Gene* was named as the most influential science book ever published, besting even Darwin's *On the Origin of Species*.

A great deal of the book's popularity is a tribute to Dawkins's gift as a communicator and his ability to take a radical scientific idea and make it widely accessible and appealing. Other evolutionary biologists, notably W. D. (Bill) Hamilton, had put forward similar ideas in academic contexts, but Dawkins seized on them, synthesizing a large amount of research to move beyond the common trope that genes are the engine of natural selection and evolution. Dawkins went further, arguing that natural selection has no interest in a single species or even an individual organism; it is interested only in the gene. Following this up he suggested that competition for the next generation would be between not individuals but genes. This may sound a bit odd, considering that there are many genes in a string of DNA, all lined up one after the other on chromosomes, and that, as we have seen, their order and number matter. Further, the process of meiosis, in which genes are swapped around, appears to be something of a lottery, with genes finding a home alongside other genes in the germ cell's new chromosomes at random. But his arguments were persuasive. Dawkins suggested that each individual gene vies to make as many copies of itself as possible. If it helps a gene's chances for replication to team up with other genes in a transient alliance, it will do so. However, he contended that, while in some instances they can work in packs, by nature genes are effectively selfish, battling each other.

One can argue with these characterizations, and many scientists did and do. But Dawkins's next idea came as a shock; I still remember the electrifying effect it had on me when, as a student, I first read it. He stated that organisms are vehicles for genes to spread, that we're disposable designs *created by the genes* with the sole purpose of moving themselves forward in time, through the organism's offspring. For Dawkins, roses, flies, slime molds, snails, condors, giraffes, humans—we're all merely devices for gene replication. Our designs and behaviors are created by the genes in their battle for the next generation.

This idea that the organism is a mere vehicle for the gene has driven the gene-centric view of life that reigns in biology today. For Dawkins, an organism is a direct, if abstract, extension of its genes. In his world, cells do not have a meaningful material basis; they do not exist in and of themselves. A hen is just a gene's way of making copies of itself, and an egg is simply the tool for achieving this by making a hen. Like many evolutionary biologists, when Dawkins talks about life, he moves seamlessly from genes to organisms to behaviors, as if they are interchangeable. Animals care for their offspring and people act altruistically toward each other because individual genes in their genomes are trying to maximize their chances of being carried forward one more generation into the future. Organisms don't behave in certain ways because it makes sense in their environment, or because that's the way they are made, or because it suits them. We behave the way we do because our selfish genes dictate it. It is a powerful idea, but by not considering the role of the cell, it buries a crucial part of the story of life.

THE SELFLESS CELL

The Selfish Gene may have been hugely popular, but scientific circles did not accept it without a fight. To defend and explain his views to fellow biologists, in 1982 Dawkins published *The Extended Phenotype*. "Once we imbibe the fundamental truth that an organism is a tool of DNA, rather than the other way around," he wrote, "the idea of selfish DNA becomes compelling, even obvious." Reading a scientist making this assertion is disorienting. Coming from a figure otherwise known for arguing the primacy of logic and scientific reasoning over belief, such a statement, for all intents and purposes, calls for a leap of faith.

To be sure, I agree with aspects of Dawkins's ideas. I concur that genes, like their material basis, DNA, are proficient replicators, just like viruses, which are for the most part DNA or RNA. And I also agree with the Darwinian canon that natural selection is competition at its core. Indeed, the portrait of our planet that Dawkins paints could easily be a world dominated entirely by viruses, which have no cellular machinery of their own and so infect and take over cells in order to replicate their genetic material and make more viruses. We've recently seen in the coronavirus (SARS-CoV-2) pandemic just how good viruses are at replicating and evolving. Insofar as DNA is a blueprint for anything, it is a blueprint for making copies of itself.

However, Dawkins neglects and simplifies much to create a picture of a gene-centered world. The copying of DNA may be important, but it requires additional machinery and dedicated space to do its work. The cell is vital to this functioning. Thus, in this world, unicellular organisms would provide the vehicles for the viral warfare. Nevertheless, a large number of organisms are made up of cells, cells are made up of proteins, and both of these entities influence what is passed on to the next generation, entirely separately from the workings of DNA. To assert that "an organism is a tool of DNA" in light of such facts is far from obvious and probably erroneous.

The fundamental logic of life defies the selfish gene thesis. If, as Dawkins suggests, life is a battle between individual genes for the prize of replicating themselves into eternity, why bother building contraptions as baroque as a eukaryotic cell? Why create the marvelous permutations of a mole's foot, a bat's wing, a porpoise's paddle, a horse's leg, and your own hand? Why create forms that require increasing amounts of energy and other resources, some with long periods between birth and sexual maturity, when they can finally serve their purpose of having offspring to carry half of their genes a generation forward? I do not like "why" questions; the law of evolution is that if it works, it will be kept. The complexity and beauty of the traits I have mentioned are obvious, and the question lurks in the background. According to Dawkins's hypothesis, genes should stick to simple single-celled options, like bacteria, or eschew the cell altogether, like viruses. Let bacteria and viruses fight a proxy war on behalf of their genes, which will be the ultimate victor so long as some cells survive for viruses to infect. Surely the appearance of eukaryotic cells can be seen as part of that game and led to new viruses adapting to them. But what is in it for the viruses in animals and plants? Single-celled organisms would be far more energy-efficient vehicles for time-traveling genes than are animals.

In limited cases, Dawkins's convictions sometimes hold up. In prokaryotes and some single-celled eukaryotes, the "selfishness" of genes makes sense. In many of these organisms, particularly prokaryotes, the DNA is, for the most part, a molecular repository dedicated to ensuring the replication of its genes. But unlike the DNA of a virus, which makes multiple copies of itself within every single cell it infects, the DNA of a cell is only replicated once, when the cell divides into two daughters. In cells, the replication of genes *as genes* is restricted. It only occurs when the volume or age of the cell leads the cell to replicate itself. When genes became components of cells, they had to abide by the terms and conditions of the cells ever afterward. Their selfishness was curtailed.

On the other hand, cells can replicate themselves independently of their DNA. During mitosis, cells use their existing structure as a template for creating a second cell. The cell's centrioles and cytoskeleton both have the ability to replicate themselves independently of the cell's DNA. So too do the cell's membrane systems. None of these parts of a cell can easily be made anew with the RNAs for their component proteins coded in the genes. They prefer to use existing structures as templates.

This independence of DNA is particularly clear in the case of membranes. Place all the genes you need to make plasma or inner membranes in a test tube, add the proteins you need to transcribe and translate those genes into enzymes, feed the contents with organic molecules for the enzymes to act on—because, after all, the enzymes need something to transform into the membrane's physical elements, and genes don't exist in a vacuum—and still a membrane will not emerge. Just as a molecule of DNA is a template for another molecule of DNA, a membrane is a template for another membrane. Cells are the entities that interpret the information in the genome and transform it into an organism. The organism is not a tool created by DNA; the DNA is the hardware store of the cell.

If we follow the argument that the cell is a replicator in the same manner that Dawkins does with the gene, we must accept that the cell, as an independent entity, is subject to natural selection too. Some evolutionary biologists acknowledge these messy facts of life by talking about "multilevel selection," meaning that natural selection can act on the level of genes, cells, organisms, and kin groups, so that each of these plays a role in what gets carried to the next generation. However, this is never explored in detail, largely because there is a big disconnect between cell biologists, who are interested in the structure and function of the cell, and evolutionary biologists, who see the world from the perspective of the gene. Most importantly, these arguments do not take into account a key difference between genes and cells: whereas genes are selfish, cells reveal themselves to be selfless.

In some instances, cells give up their individuality to provide a function necessary for the organism to survive. In an extreme example, during the building of our muscles, individual muscle cells fuse with each other to generate large fibers with multiple nuclei and mitochondria. This ensures that they can generate the energy and maintain the power and strength needed to sustain the strains of physical activity performed as part of everyday life.

Such cooperative acts lead to new levels of emergence. For instance, our thoughts, feelings, and movements are the product of signals relayed between neurons. Our neurons are not merely operating individually, however; they're connected in large functional units called circuits, which govern neural functions. Neural circuits have organized in various ways. Some diverge, where a signal from one neuron or small group of neurons stimulates many other neurons, perhaps thousands. Some converge, where signals from many neurons stimulate just one neuron or a small group of neurons. Some stimulate chains of neurons, creating parallel or reverberating signals. From these circuits emerge our mind and consciousness. While genes provide instructions for the proteins needed to build these circuits, it is the cells that interpret ensembles of proteins to hook up with each other in these different ways. The circuits are the products of interactions among cells, not genes, and their output is another, well-accepted example of emergence.

The life of an organism can be marred by the selfishness of genes. Occasionally, a gene or two breaks loose from the control of the cell. When the products of such genes are involved in cell division or communication, they do what they do best: replicate, mutate, gamble their future by creating variation. When this selfishness gets out of control, cancer ensues. Cancer is sometimes referred to as a disease of "rebel cells." Though it affects cells, cancer is not a disease of cells; it's a disease of genes, the result of genes rebelling against the control of the cell, rising up against their cellular masters to force the cells to replicate at the genes' beck and call. When genes replicate and mutate without anything to stop them, they destroy the organism—just as a virus will do when immune system cells fail to stop it.

Taking a cell's-eye view of life reveals a tug-of-war happening within multicellular organisms. The eukaryotic cell's impulse is to associate and cooperate with other cells, to build organs and tissues and, through them, an organism. The gene's interest is to replicate itself ad infinitum. The wealth of diversity in nature, most especially across animals, suggests that cells and genes must have formed some sort of entente to resolve their cross-purposes.

A FATEFUL PACT

When the first eukaryotic cells formed from the merger of some bacteria and archaea, something shifted in how genes pursued their ambition of eternal replication. In the first place, two or more genomes had to come to an agreement between themselves: in order to survive, they had to temper their selfishness. One genome settled into the mitochondria, another into the nucleus. But more than this, the genomes had to temper their selfishness with respect to their host, the cell, because the cell was only able to carry more than one genome if it also had a say in its own structure and organization. In fact, over the course of evolutionary time, some of the bacterial genes that gave rise to the mitochondria moved to the nucleus. When multicellular cooperation became attractive to cells, such pacts became paramount, rewriting the contract between genes and cells.

As we have seen, the emergence of animals some six hundred million years ago coincided with an extraordinary expansion in the number of protein-coding genes contained in the genome. Many of these newer tools specifically helped (and continue to help) cells expand their interactions with other cells and the environment, change their shape, and focus their functions. Cells could also use them to come together to build an organism. Having access to this vast catalogue of hardware unleashed the inventive potential of cells, allowing them to gather and grow into the diversity of animals, fungi, and plants. The creativity of cells and the immortality of genes became entangled.

In Goethe's classic play *Faust*, the dedicated and godly scientist Dr. Faust is lured into a contract with the Devil. In return for a moment of transcendence in life, the doctor hands over his soul to the Devil for eternity, gaining short-term brilliance and fame at a long-term cost. I view animal cells as having struck a similarly Faustian bargain with the genome. In this pact, the genome—playing the Devil—allows cells to use its coding potential to build and maintain a complex, multicellular organism for the length of its life, so long as the genes will be passed on, blamelessly, to any subsequent generations. This is the role assigned to the germ cells and gametes, which physically carry genes to the next generation. These cells are set aside, more often than not, before an animal starts to be built.

Building animals and plants requires the creation, arrangement, differentiation, and coordinated activity of many cells and cell types. These arise after the germ cells have been set aside in a self-contained area of the organism. The germ cells have no input into how the organism is built. Indeed, during the only time that germ cells make use of the genome, they use hardware dedicated to protecting DNA from being altered and performing meiosis. The function of the germ cells is to ensure that a pristine copy of the genome is sheltered from the hustle and bustle of animal construction. And, as a matter of fact, as soon as the germ cells emerge, they shut down forever the possibility of becoming any other cell; they close down the genetic programs that lead to the generation of different cells. This is the hard part of the pact for the cell. In the rest of the organism, though, cells can do with the genome what they will. So while every other type of cell in the body harnesses the genome to further multicellular cooperation, the germ cells and gametes are harnessed by the selfish genome, which uses these special cells as their vehicle for time travel, untouched by the creative processes associated with the generation of the organism. From the point of view of the cell and the organism, the genome is a toolbox that cells use to make a hen, and the egg is simply the hen's payment for accessing it.

While genes help us to understand the process of natural selection, they do not, on their own, explain how a fin evolves into a paddle, hand, foot, or wing. Mutations in genes introduce changes in the hardware catalogue, opening up new creative possibilities, but it is the cell that decides which new tools are kept and used and which are tossed out. Before natural selection, the cell makes a selection of its own.

Darwin came close to grasping this very insight. In chapter 13 of his great book, he edged toward the intuition that the working of cells might have driven the emergence of the variety of vertebrate forms from a simple, common starting point. He does not say so explicitly, but he does note that embryos of many species look eerily similar in the early stages of development, only displaying their endless differences later on. He was on to something. The first structures that cells create, embryos, give us a vital glimpse of the origin of animal diversity. Long before DNA could reveal the shared ancestry of cells, embryos provided evidence of organisms' "descent with modification" from a common progenitor.

Embryos are a beautiful example of the craftsmanship of cells. As we shall see, they are a continuous act of emergence in which volume, form, function, and time come together to create the beauty of what we call an organism. This process is the consequence of the pact between the gene and the cell and lies at the root not only of the development of the organism but of the rich variety of form and function that we admire around us.

PART II
THE CELL AND THE EMBRYO

I awaited in excitement the picture which was to present itself in my dishes the next day. I must confess that the idea of a free swimming hemisphere or a half gastrula with its archenteron open lengthwise seemed rather extraordinary. I thought the formation would probably die. Instead the next morning I found in their respective dishes typical, actively swimming larvae of half the size.

—HANS DRIESCH, "DER WERTH DER BEIDEN ERSTEN FURCHUNGSZELLEN IN DER ECHINODERMENENTWICKLUNG"

Thus, as it seems to me, the leading facts in embryology, which are second to none in importance, are explained on the principle of variations in the many descendants from one ancient progenitor, having appeared at a not very early period of life, and having been inherited at a corresponding period. Embryology rises greatly in interest, when we look at the embryo as a picture, more or less obscured, of the progenitor, either in its adult or larval state, of all the members of the same great class.

—CHARLES DARWIN, ON THE ORIGIN OF SPECIES

Finally, therefore, we may state the conclusion that all the cells or their equivalents in the fully developed organism have arisen by a progressive segmentation of the egg-cell into morphologically similar elements; and that the cells which form the early basis of any part of an organ of the embryo, however small their number, form the exclusive source of all the formed elements (i.e., cells) of which the developed organ consists.

—ROBERT REMAK, *UNTERSUCHUNGEN ÜBER DIE ENTWICKELUNG DER WIRBELTHIERE*, 1855

– four –

REBIRTHS AND RESURRECTIONS

MBRYOS ARE SMALL STRUCTURES—FROM 0.5 TO 1 MILLIME-
ter long, depending on the species—made up of thousands of
cells that, in a rich display of diversity, presage the range of different
cell types that will configure the organism. The number and precise
organization of these different cells beg the question of how they
come about from the one cell, the original zygote. The diversity is
enormous, and we have seen examples of it in the nervous system,
but it is the same in other parts of our body.

Even within a single layer of a single structure, cells exhibit
startling variety. A palisade of cells—a layer of cells stacked in an
organized fashion—spans the length of your digestive tract, but
even if we tighten our focus on just this palisade in just this tract,
we will witness remarkable variety in cell type and function. Those
cells near the beginning of the tract, in the esophagus, are stacked
in thick layers to provide a barrier of protection against incoming,
undigested food. By the time that food gets to your stomach, this
palisade has become a single layer of diverse cell types, some of
which secrete acids and enzymes to break down food, while others
secrete hormones to regulate the stomach's activities and defend
against bacterial threats. Still lower, a single layer of cells lines the

duodenum, the first part of the small intestine. These cells absorb the nutrient particles from digested food, while others secrete a different set of hormones, including those that trigger postprandial drowsiness or protect your intestine from acidic juices that might leak from your stomach. The society of cells is not homogenous, and this creates opportunities for cooperation with functional consequences. However, over the last few years we have discovered that even within the same organ or tissue, cells deploy a wealth of genetic activity, what we could call variations on a genetic theme.

All of these cells ultimately come from your very first cell, the zygote. How that single cell produces such diversity, how cells choose which genes to use, when and where, and which to avoid were central questions in biology long before anyone had coined the term *gene*, let alone sequenced a genome. Some answers began to emerge from the first successful artificial cloning experiment, conducted on a beach in the Bay of Naples back in 1891.

Hans Driesch was a young biologist spending his summer at the Stazione Zoologica, an institution set up near the shoreline for the benefit of researchers from around the world studying the development and physiology of marine organisms. The station remains a hub of scientific innovation and exchange to this day.

Curious to see when and how the cells of an animal diverge from each other, Driesch aimed to test a hypothesis popular at the time holding that each type of cell was built using specific elements, called *determinants*, which were in turn contained in the original zygote. As cells divided and multiplied, the claim went, they only got a portion of the original kit of determinants, and the portion they got determined whether the resulting cell gave rise to a part of an eye, an arm, or some other organ or tissue of the body. This thought had gained support after a series of experiments conducted in the 1880s by Wilhelm Roux. Roux had fertilized eggs of the green frog (*Rana esculenta*) and waited for the zygote to go through its first division to create two cells. Then he had poked one of the two cells with a hot needle to kill it. As the remaining cell multiplied and developed, he observed that it looked like a "half embryo," a lopsided

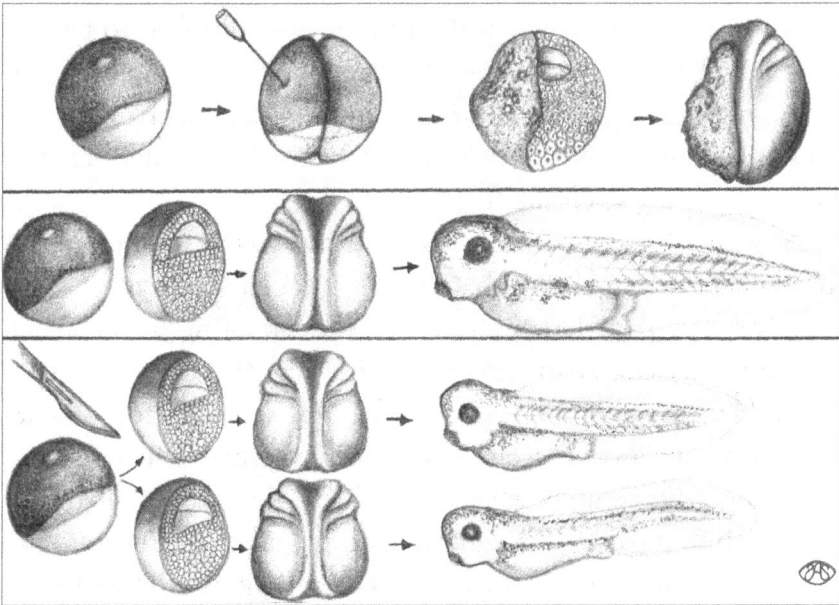

FIGURE 14. *Top:* the lopsided "half embryo" Wilhelm Roux observed after killing one of the two cells following a frog zygote's first division. *Middle:* normal development. *Bottom:* the result of separating the two halves, which leads to two normally patterned but half-size embryos.

frog embryo in which only one side of the body was developing as expected. From this observation, he concluded that even though a zygote can give rise to a whole organism, subsequent cells do not possess the same animal-building potential as the initial progenitor.

Driesch did not like this idea. He thought something in Roux's experimental design had biased the results. So he repeated the experiments, but rather than using frogs—large and complicated animals—he chose as his subject the sea urchin, abundant in the Bay of Naples right at the threshold of his lab. Also, instead of killing cells that remained attached to the rest of the developing cluster of cells, he sought to gently tease the cells apart and see how they developed independently. "I awaited in excitement the picture which was to present itself in my dishes the next day," he wrote. "I thought the formation would probably die." But when he arrived at his lab the next day, he found something "rather extraordinary":

two complete, actively swimming sea urchin larvae. Twins. He repeated his experiment when the zygote had multiplied to be four cells and obtained four larvae. At some point the magic was lost: he was not able to produce infinite urchins from more than the four cells of an embryo. Nevertheless, he had struck a resounding blow to the prevailing scientific consensus.

As with all good science, it was necessary to test the new results in other species to ensure this wasn't a peculiarity of sea urchins. Others followed in Driesch's footsteps by adopting his approach with frogs, teasing apart the cells and letting them develop independently, producing the same result: two complete frogs. Something had gone awry in Roux's experiment. Perhaps leaving a dead cell among the group of early cells, which came to be called *blastomeres*, had affected the embryo's development. Scientists also started to question whether Roux had been right in calling what he saw a "half embryo."

Over a century later, Driesch's findings have held up. When the first two cells of a developing rabbit or mouse are separated, each of them gives rise to a complete animal, not two halves. It's very likely that many identical twins are the result of a similar, but accidental and spontaneous, separation of early cells in the womb. While such an event is rare in humans—only one in four hundred births—in some other animals it is something of a reproductive habit. For example, the nine-banded armadillo is particularly fond of twinning, producing not two but four embryos out of every single fertilized egg.

As Driesch recognized at the time, the implications of this observation are extraordinary. Each cell in the very early embryo has the potential to give rise to a whole organism. To put it another way, they are *totipotent*, a word whose roots mean *whole* and *effective* or *powerful*. We can think of totipotency as a sort of insurance policy against the loss of cells at a time when every cell counts. This period varies from animal to animal, but it is usually brief: in the case of the sea urchin, by the time there are eight cells, the cells have differentiated; in mammals totipotency lasts longer.

There's a curious twist to totipotency. If, instead of teasing apart the blastomeres, you fuse early cells from different zygotes, the blastomeres combine seamlessly to produce a single organism with two genomes. Say you take two groups of cells, one of which would have developed into a white mouse and the other into a black mouse. Fuse them in the early stages of development, and you'll get a perfectly proportioned, spotty-stripy mouse, with the same number of cells as a typical mouse.

These cell fusions create chimeras, like Karen Keegan, whom we met in the introduction. Chimeras offer another example of cells' ability to count, revealing that they are able to regulate how many cells are in a tissue, organ, or organism at a time, sensing their relationship to the cells around them, choosing which genes to use. In fact, chimeras have been made between a sheep and a goat with surprisingly beautiful results.[1]

FROGS FOREVER

The French biologist and philosopher Jean Rostand is often remembered for his saying *Le biologist passe, la grenouille reste*: the biologist passes away, the frog remains. Such is the fate of efforts to understand living systems. The frog offers a perfect exhibit for representing the complexity of life not only because of its humble status in the animal kingdom but also because it has served as one of the primary research subjects for understanding how cells become different from each other.

There are many reasons why frogs have dominated laboratories and biology textbooks. A frog's egg is about 1.2 to 1.5 millimeters in diameter—huge compared to the puny 0.15 millimeter or so of the mammalian egg. Most of the space inside the egg is occupied by yolk food for the cells that start multiplying after fertilization; notice that the embryo will emerge from the egg, but it is not the egg. As the cells consume the yolk, space opens up, and they can even more easily be observed and manipulated within the eggshell. On top of that, despite the errors of Roux's needling, frog zygotes are quite robust, capable of surviving and healing themselves after surgical interventions. You can remove cells, add cells, and exchange their positions, and most of the time an embryo will still develop into a larva. Frogs also develop quickly; it takes over fifty hours for a zygote of the African clawed frog (*Xenopus laevis*), a favorite of embryologists, to become a tadpole. Altogether, they make an almost ideal experimental subject for the study of development.

Frog eggs are particularly ideal for examining the limits of totipotency. Scientists interested in determining when and how cells lose their ability to give rise to a whole animal found the large size of the frog egg enormously helpful in conducting experiments. Given that Driesch had seen that totipotency is lost at some point, scientists following in his tracks wondered about whether it could possibly be recovered. Is the ability to give rise to other cell types lost after cells differentiate into neurons, muscle, and skin? Or do cells, rather than throwing away unneeded tools, stow them somewhere safe for later use? For those asking such questions, the frog's egg readily provided answers.

In 1952, a year before Rosalind Franklin, Francis Crick, and James Watson proposed their double helix model of DNA, two American scientists working with the northern leopard frog (*Rana pipiens*) began to interrogate the capabilities of cells during development. Robert Briggs and Thomas J. King, at the US Institute for Cancer Research in Philadelphia, took an *oocyte*, an unfertilized egg, removed its nucleus, and implanted the nucleus of a cell that had reached a more advanced stage of development. Having ensured that the egg now possessed the number and complement of chromosomes it would have after fertilization, they tricked it into believing that a sperm had actually fertilized it and watched what happened. They hypothesized that the manufactured zygote would not develop very far; it might not even make an embryo. This would help confirm that cells lose some or all of their developmental potential once they have differentiated.

The experiment was easier said than done and required a huge amount of skill on the part of the researchers. Each egg had to

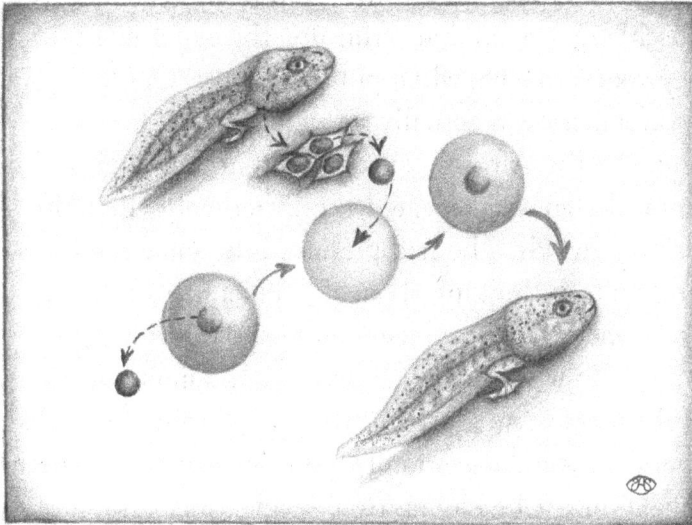

FIGURE 15. The concept of cloning by transplanting a donor nucleus into an oocyte pioneered by researchers Robert Briggs and Thomas J. King and performed by John Gurdon in *Xenopus*.

be penetrated gently to remove its nucleus with minimal disruption to the rest of the cell. The donor cell nucleus, taken from a differentiated cell, was then inserted using a syringe whose barrel was slightly narrower than the cell's diameter, and the newly nucleated egg was fertilized. This required a great deal of patience. Once Briggs and King succeeded in doing this, their results were unambiguous: when they swapped in nuclei from the cells of early embryos, their eggs produced some tadpoles. When nuclei came from older cells, they got hardly any. The process of development, from zygote to blastomere to embryo to adult, seemed to be irreversible at the cellular level. After differentiation, it seemed, cells put their bag of tricks away.

We scientists tend to be skeptical by nature, however. The fact that something does not happen in one experiment does not mean that it does not happen at all. As the saying goes, absence of evidence is not evidence of absence. One of the skeptics in this case was University of Oxford student John Gurdon. He felt that Briggs and King's experiment was so technically challenging that something could have gone amiss, and his academic advisor, Michael Fischberg, agreed that rerunning the experiment would be a worthy project. Maybe, when put to the test, Gurdon would find that development was actually reversible after all.

Turning to the African clawed frog, Gurdon improved the experimental design and repeated the experiment. In 1956, shortly after starting his PhD, he had produced the same results as Briggs and King. But, in thinking about the difficulties of the process, he began to wonder if the problem had less to do with differentiation and more to do with time. Cell division and multiplication in the early embryo is fast and furious, whereas differentiated cells divide slowly, and in some cases rarely. Gurdon wondered whether the older nuclei might be having trouble adapting to the pace of cell division in the earliest days of development. It might be hard work for a nucleus to catch up to the rapid pace of cell division if it's just been awakened from a long sleep.

To give the nuclei from differentiated cells a chance to adjust to the pace of early development, Gurdon introduced a number of technical innovations that facilitated the procedure and invented a new technique called *serial transplantation*. He took the nucleus of an older, differentiated cell, inserted it into a younger, undifferentiated cell, and allowed that younger cell to divide a number of times. Each of the resulting embryonic nuclei was then inserted individually into a new oocyte following Briggs and King's original concept. Since the nuclei were all descendants of the same differentiated cell, the new procedure should not make any difference to their potency: if the genes that give cells the power of totipotency are silenced or erased forever when cells differentiate, these descendant nuclei wouldn't produce a whole organism. If Gurdon was correct, however, and the problem was a mismatch in the pace of cellular activity, serial transplantation would give the older nuclei a better chance of assimilating to the speed of youth.

As it happened, when he used donor nuclei from blastomeres and early embryos, Gurdon got a slight improvement in the number of embryos that developed into tadpoles but largely obtained the same results as Briggs and King. However, when he used donor nuclei from the intestines of tadpoles, the experiment produced a host of swimming tadpoles that developed into fertile frogs.[2] Gurdon was inserting even older, more differentiated cells than the earlier experiments had used; yet he still found that the capacity for totipotency could be retrieved. He repeated the experiment with nuclei from a range of cell types from tadpoles and continued to produce swimming tadpoles that developed into fertile frogs. The process of cell differentiation did not alter the information in the chromosomes. All of the information was still there and could be retrieved—although it might require some work. In a further set of experiments, he used nuclei from the tissue of fully developed frogs; this never worked. Nonetheless, the point was made: cell differentiation was reversible; youth, too, might be retrievable.

Gurdon's experiment not only had shown that the cell's hardware store, its toolbox, remains intact after differentiation but had also generated the first hint that cloning an animal using a cell from a fully developed adult was feasible. But despite his remarkable discovery, Gurdon's research did not arouse much concern or interest among the general public. Perhaps the science was too technical and abstract to engage people in a year when the space race was at full tilt, and Francis Crick was promulgating his central dogma of molecular biology: that once information has passed from DNA into protein, it cannot get out again.

Or perhaps frogs are too far removed on the tree of life from humans. Perhaps it was just a matter of selecting a cuddlier, woolier, farmyard animal to capture the public's interest in the incredible power of zygotes and the worrying ramifications of the advent of cloning.

FROM DOLLY THE SHEEP TO CC THE CAT

Gurdon's frog experiments may have done much to demonstrate that cloning was possible, but they left many scientific questions unanswered and raised new ones. Some scientists wondered whether the ability to "wake up" the nucleus was peculiar to amphibians; others didn't believe the results at all. A common criticism holds that Gurdon never succeeded with an adult cell, forgetting that intestinal cells from a tadpole are fully differentiated and functional.

It might seem that the next step would be to replicate the injection of older nuclei into mammalian embryos, but the process was fraught with obstacles far beyond those conquered by Briggs, King, and Gurdon. For one thing, mammalian eggs are, on average, less than 0.1 percent of the volume of a frog's, which created a big challenge for the needle poking required by the experiments. Assuming that scientists could find a way to successfully transfer a nucleus into a space a thousand times smaller than a frog's egg, they would still need to work out the next step: implanting the egg with the transferred nucleus into a surrogate mother's womb. To accomplish this, they would need a womb primed to receive the embryo and carry it to term. Given that only 1 to 5 percent of cloned frog embryos developed to adulthood, this meant that researchers would potentially need hundreds of fertile female mammals at the ready to receive an embryo. It was daunting, but this did not stop scientists from trying.

The hunt was on, and researchers were soon trumpeting results. In 1981, Karl Illmensee, a biologist at the University of Geneva, announced that he had successfully cloned a mammal—three mice, to be exact—by transferring the nucleus from an early embryonic cell into a mature egg. However, doubt was soon cast on his results, which were found to be the product of dark craftsmanship rather than reproducible science. In fact a true mouse clone wouldn't be born for almost two decades. In the meantime, another line of experiments tested the capacity of bacteria to serve as protein and hormone factories. Oddly enough, this line of research would end up paving the way for the world's most famous mammalian clone.

These efforts were initially less intent on creating clones than on producing useful hormones. Science often works in oblique ways. In cataloguing the genome, scientists recognized that cells take advantage of a kind of index listing the possible proteins, hormones, enzymes, and other body-building hardware at their disposal that is written in the transcriptional control regions responsible for determining what genes should be expressed or silenced and under what circumstances. Once it became clear that the genome is agnostic as to which messenger RNA (mRNA) is going to be produced, scientists realized they could insert a gene with the code for a specific protein or hormone into a bacterium, under an indiscriminate bacterial control region, and use it to force the bacterium to express that gene. The bacterium would dutifully respond by transcribing its DNA into mRNA and translating its RNA into the desired protein or hormone. In this manner, researchers were able to convert bacteria into hormone factories, efficiently producing large amounts of insulin and growth hormone and making the older, laborious biochemical processes obsolete.

Still, there were some proteins that bacteria simply would not make, at least not in any useful form. This is because some proteins require chemical modifications that only their native mammalian cells can provide. For example, one protein that bacteria cannot make in any useful form is Factor IX, which is essential to blood clotting and the treatment of hemophilia. No matter how hard scientists tried, they could not force bacteria to make a functional version of this protein.

For quite some time the idea was bandied about that a solution to this problem might come from mass-producing these finnicky proteins in the milk of sheep and cows. Applying lessons from the bacteria protein factories, it would be possible to hijack a control region governing expression of a milk protein in mammary glands and use it to, instead, express any protein, for example, Factor IX; the mammary gland cells should be able to introduce those modifications that make the protein active. Such bioengineering would create *transgenic* animals, creatures into which DNA from another organism has been introduced.

In 1989 this theoretical process became a reality. John Clark, a molecular biologist at the Edinburgh Research Station of the

Institute of Animal Physiology and Genetics Research, succeeded in making genetic modifications to a ewe so that her mammary cells produced human Factor IX in her milk. Clark also managed to modify a second ewe, called Tracy, to produce in her milk Alpha-1-antitripsin, a human protein used in the treatment of emphysema and cystic fibrosis. The process was not easy. Moreover, because of the jumbling of paternal and maternal genes during meiosis, there was no guarantee that the offspring of a transgenic ewe would produce the same human proteins in the same quantity, if at all. The mammary gland cells of a ewe would never be 100 percent identical to its mother's—unless, of course, the offspring were clones.

The challenge was now to remove the genetic lottery of reproduction by cloning the most productive transgenic ewes. To accomplish this task, the Edinburgh Research Station evolved into the Roslin Institute and a spin-off company, PPL Therapeutics, with the express goal of cloning animals for commercial production of biomedically useful proteins. The researchers called it "pharming," the conjoining of pharmaceuticals and farming. In 1995, a joint Roslin-PPL team had produced twin ewes, Megan and Morag, from embryonic cells, extending the work of John Gurdon into mammals. However, whether they could do the same with adult cells and produce transgenic sheep at scale remained an open question.

From 1995 into 1996, a team under the leadership of Ian Wilmut set about transferring embryonic and fetal cell nuclei into oocytes in an effort to refine the process. Such work demanded that they make the most of female sheep's fertility during the breeding season. One day in February 1996, they faced a setback when none of the cells they had planned to use for transfer were viable. "The last thing you want to do is waste those oocytes," Karen Walker, then an embryologist at PPL, later recalled. "We wanted to try something, at least." They looked around the lab and found some udder cells from a six-year-old, "middle-aged" ewe that had been prepared by Keith Campbell, a talented cell biologist who had been working with Wilmut to identify conditions that might allow the addition of adult nuclei into the rush of early development. Even though they had not yet identified these conditions, they decided to make the most of the breeding season and try them. Nothing ventured, nothing gained.

Five months later, on July 5, 1996, Dolly the sheep was born. She was the first mammal—and arguably, the first animal—ever cloned from an adult cell.[3] As Gurdon had hypothesized but never succeeded in proving with frogs, a fully differentiated adult cell nucleus could be reactivated so that it once again became totipotent. During development, cells select specific genes to build specific tissues with specific functions. Those that are not needed may be put aside, but though they are typically inaccessible, Dolly proved that they were recoverable after all.

Dolly became an instant celebrity. Her remarkable birth ushered in a brave new world, though it was not clear to the public if this world signaled promise and excitement or a dark, dystopian future. Some suggested that the Roslin Institute had kept Dolly secret for several months because it was planning to clone humans. The team categorically denied claims of this type; this was a milestone in developmental biology to be sure, but the scientists' goal remained to create herds of transgenic sheep clones for use in producing medicines. The following year, two more ewes, Polly and Molly, were born using adult cells that had been reengineered to produce Factor IX in their milk. The process was difficult, but it worked.

No amount of denial from the scientists could stifle the growing trepidation about human cloning. Responding to popular anxieties, in 2005 the United Nations declared a nonbinding moratorium against human cloning, "inasmuch as [it is] incompatible with human dignity and protection of life." More than forty countries have since passed laws banning cloning of humans, although some allow clonal embryos for use in research—the obvious loophole in the vaguely worded UN ban. But regardless of whether human cloning is allowed, the creation of a true doppelganger remains a technologically distant possibility. That's because, as scientists have grown more adept at generating animal clones, the offspring have revealed a few surprises—including that even a clone is not actually an exact replica of its original.

In 2001 the first pet to be cloned was born. Her name was CC, for "Copy Cat," and although she shared all her genetic material with her donor parent, a cat called Rainbow, the two cats' coats were very different. CC was a tiger tabby, all browns and white, while Rainbow was a calico, with big blocks of colored fur. This is because many of the genes involved in coat color reside on the X chromosome. Rainbow, as a female cat, had two X chromosomes. Each pigment-producing skin cell only needed to use one of the two genes on the X chromosomes and turned the other off, but this happened cell by cell. CC was cloned from the nucleus of a cell that had deactivated either the "orange" or the "black" gene, and it no longer could gain access to those particular tools. Similar gene inactivation phenomena can be seen in identical twins. Although they have the same DNA, they exhibit many differences, some quite subtle. In a dramatic case, this random inactivation of a gene can lead to surprising differences in monozygotic twins. The gene *DMD*, associated with Duchenne muscular dystrophy, sits on the X chromosome. In females, one of the two chromosomes is inactivated so that all cells have the same number of genes in that chromosome as a male, which only has one X chromosome. This inactivation is random and happens on a cell-by-cell basis. In many instances in which two twins are heterozygotes for a mutation in *DMD*, often one may develop the disease whereas the other won't, depending on which chromosome is inactivated in which cells—even though both share the same genome. Another reminder that an organism's uniqueness is found not in its genes but in how cells use them.

The company that produced CC, Genetic Savings & Clone, was set up as a commercial enterprise to exploit the new science of cloning, but after just five years in business, it shut down. Its inability to duplicate a donor was a real problem. People didn't want a cat with the same genes as their pet; they wanted a cat that looked and behaved exactly the same as their pet. They wanted to revive their dearly departed pet's cells, not its genes, and that's simply not possible.

Commercial cloning in general has turned out to be more of a pipe dream than a reality. PPL Therapeutics is no more. The company wound down its operations and sold its assets in November 2003 after investors grew impatient with the slow pace and high cost of developing pharming products. Nonetheless, the cloning techniques they pioneered have been adopted in some sectors. Cattle breeders use them to maintain animals with high yields of milk. Other breeders took up the somewhat whimsical goal of producing and duplicating the perfect Texas steak. Racing horse breeders have also taken an interest. A guaranteed Derby winner has the potential to earn an enormous amount of money, and so even a cloning success rate of about 10 percent could be worth the expense.

More important, in many ways, are the potential noncommercial applications of cloning, which can aid in the preservation of biodiversity. Though the resurrection of extinct species through cloning is, at least for now, impossible since you need living cells to create living animals, it could be practicable to pursue the use of cloning to preserve species that are close to extinction or face reproductive challenges. In the case of the Pyrenean ibex (*Capra pyrenaica pyrenaica*), researchers managed to catch the last female of the species in 1999. They named her Celia and collected skin cells from her, which they froze in liquid nitrogen prior to her death a year later. Emboldened by Dolly, Polly, Molly, and other clones, José Folch and his colleagues decided in 2003 to clone Celia from her frozen hide. Lacking Pyrenean ibex oocytes or surrogate mothers, the scientists turned instead to close surviving relatives: nuclei from Celia's skin cells were transferred into domestic goat eggs, and then, where the fusion succeeded, an early embryo was implanted to form a hybrid zygote of a Spanish ibex and a goat. Out of 782 embryos created in the experiment, 57 lasted long enough for implantation in a surrogate mother, and 7 embryos attached to the wall of the womb to create a connection to the mother. Only 1 of those 7 embryos survived to birth, but that newborn died within a few minutes of her delivery by caesarean section. She had developed with a fatal defect in one lung. Still, the clone's birth was hailed in the media as the first "de-extinction" event. The arithmetic was challenging—especially given that in order to truly resurrect a species, you would need not only to nurture a healthy embryo to adulthood against steep odds but to do it at least twice, creating one male and one female capable of surviving long enough to reproduce themselves.

Scientists were undaunted by the odds. Building on the approach used in the ibex cloning experiment, in December 2020 a black-footed ferret (*Mustela nigripes*) was cloned from skin cells of an animal that had been dead for nearly forty years. Black-footed ferrets are critically endangered, according to the International Union for Conservation of Nature, particularly because the one thousand or so animals surviving in the wild are so genetically similar. The hope is that the newly cloned arrival, called Elizabeth Ann, will help to restore diversity to the toolbox available to the species. The research team chose the donor cells for Elizabeth Ann and her descendants thinking that they might be better able to fight off sylvatic plague—a disease that has decimated the ferrets and is caused by the same bacterium as bubonic plague in humans.

BRING BACK THE MAMMOTH?

The popular fascination with resurrecting extinct species took its most famous form in Michael Crichton's *Jurassic Park*, a novel in which scientists find a way to bring back dinosaurs with, as one might expect, less than idyllic results. Yet there are some who regard *Jurassic Park* not as science fiction but rather as a challenge.

Renowned and maverick genome scientist George Church of Harvard University and entrepreneur Ben Lamm have set out to bring the woolly mammoth (*Mammuthus primigenius*) back to life through cloning. Church is not the only scientist aiming at this feat—mammoths seem to be an obsession in the de-extinction field—but he has obtained enormous backing. Founders of the tech start-up Colossal Laboratories & Biosciences, Church and Lamm believe that this accomplishment could usher in a new genetics that will not only turn *Jurassic Park* into reality but also allow us to create entirely new life-forms to do our bidding: remove methane from the Arctic or clean our oceans of plastic.

These lofty dreams run aground on a difficult reality: it's not possible to create any life at all through genes alone. As Rudolf Virchow, founder of the cellular view of disease, famously said, "Where a cell arises, there a cell must have previously existed (*omnis cellula e cellula*)." All cells from cells.

To truly resurrect a mammoth via cloning, Colossal would need to start out with mammoth DNA in hand. Researchers have identified some pieces of this DNA from fossils, and from these they *might* be able to synthesize large stretches of the mammoth genes they will need for their experiments. However, even if that task were successful, they would next need to put that DNA into a living mammoth cell, preferably an egg. Unfortunately, all mammoth cells died when the species went extinct, and it's notoriously difficult to wake up a cell from the dead. Lacking a mammoth egg, they are opting instead to insert the genes into a recipient cell from a relatively close species, like the Asian elephant (*Elephas maximus*). However, given that they do not have, and cannot insert, an entire intact mammoth genome into such an egg, this path too is a dead end. So it's on to option 3, a more humble experimental design that involves "tweaking" an Asian elephant genome to make it resemble a mammoth's.

This is exactly what Church has set out to do. He and his team have indicated that they intend to start by choosing a few genes associated with specific and essential features of a mammoth. They'll take an elephant cell and, using CRISPR technology—the set of so-called genetic scissors that bacteria use to detect and destroy viruses that their genome has previously encountered—snip out specific elephant genes and replace them with their mammoth counterparts. Then they will extract their "engineered" nucleus and place it in an Asian elephant oocyte, or immature egg, whose own nucleus has been removed, to create a hybrid zygote. In case this doesn't work, they have more advanced techniques to reprogram cells, which we shall discuss later. This zygote will next be chemically stimulated to start growing, after which it will be implanted in a surrogate elephant mother for gestation. If all of this goes well, they will possess a creature that is not quite an Asian elephant and not quite a mammoth either.

It is unlikely that such a process will go well in the first place. For now, let's set aside the fact that IVF in elephants is, both literally and figuratively, an embryonic field and there is a strong chance any single pregnancy will fail. Assuming that such an implantation is successful, when the miraculous baby is born two years later, the animal that emerges into the world will be a marvel—but not a mammoth. It will likely have different hair and different teeth from a modern Asian elephant, as per its design. Its red blood cells may unload oxygen more easily because they will have a special type of hemoglobin that is less sensitive to cold conditions. It may look and live somewhat like a mammoth, as far as we can visualize a creature that went extinct long ago. But most of its DNA and all of its cells will still be those of an Asian elephant. At best, Colossal's creation will be nothing more than a "mammouphant," and then, don't forget, it will have to breed.

I am among the scientists who are not persuaded that this bioengineering experiment will work, most certainly not within the six-year time frame that Church has suggested. But Church's experimental design does highlight the secondary role that DNA plays in these processes.

Like Richard Dawkins, Church and his fellow resurrectionists cast the cell as a passive vessel for mammoth DNA. They insist that bringing an extinct species back to life boils down to creatively rewriting and re-creating genetic code. And yet the plans for reviving mammoths are rife with cells. From harvesting a donor cell, to inserting an edited nucleus into another oocyte, to implanting a mature pseudo-zygote into a surrogate mother (herself a conglomeration of cells), this is a story about the power of the cell, not the gene. If building an embryo only takes the right genes, why bother with the difficulties of working with an elephant egg in the first place? A reprogrammed nucleus should do this job no matter what cell it found itself in—that is, if the genome were as powerful as Church and Colossal would like to believe. But it is not. You cannot put DNA in a test tube and expect an organism to emerge. The stubborn reality is that, without a cell—the *right* cell—a genome is a string of As, Gs, Cs, and Ts. When Gurdon or the scientists at Roslin place the nucleus from an adult cell into an oocyte, it is the goings-on in that cell that perform the magic of rejuvenating the incoming genome.

This raises the question of how the cell reaches out to the genome to select the genes it needs to adopt an identity, a shape, a purpose during development. Of course, the same question applies to the magic of the oocyte. The basic ideas and rules derive from a surprisingly simple and different world, that of bacteria, and the insights of two French researchers playing with mutants.

THE BUG AND THE ELEPHANT

Cloning experiments demonstrate that the genome remains pretty much the same in all cells across the course of the organism's lifetime. Far more important is the fact that as development proceeds, cells pick and choose the genes they need to create this or that tissue, hiding away what they don't need. But how does a cell know which hammer or hinge it needs for the particular type of cell it's building among the two hundred or more cell types that make up a human? How does it make sure that the tools it doesn't need are packed away safely so they don't interfere with their regular functions? In other words, how does it know what type of cell it is? Researchers found answers to these questions in a series of experiments involving a simple form of life, far removed from the majesty of sheep or humans, but with its own story to tell.

Bacteria are, at first sight, boring organisms whose lives amount to little more than looking for food, reproducing, and protecting themselves from viruses. Our bodies are teeming with about one hundred trillion bacteria amid our forty trillion human cells, though some contribute to our existence in beneficial ways, including those that are essential to digestion. Our internal balance can be thrown off, however, when harmful bacteria arrive from the external world—who has not had a "stomach bug" at some point?—or when antibiotics wipe out helpful bacteria, disrupting our bodily functions.

Under normal circumstances, the bacteria in your body rarely face existential challenges, on the scale of your cells, tasked with deciding whether they should become a muscle cell or a nerve cell, multiply or differentiate, live or commit suicide, as some do. However, this doesn't mean that bacteria lack choices. Most of their choices concern what and when to eat, and as it turns out, how they pick their menu reveals general principles about how cells make choices about a whole host of things.

Escherichia coli, or *E. coli*, as it is often known, is one of the bacteria that lives in our gut. It scavenges the by-products of our own cells' digestion of food, breaking down the materials further into nutrients that our cells can absorb and use as fuel. By observing "domesticated" *E. coli* in controlled conditions in a lab, scientists have been able to discern the bacterium's economical approach: when presented with a buffet of food choices, it dines on the options in ascending order of the amount of effort eating them takes. For example, when served a simple choice of two sugars, glucose and lactose, *E. coli* will invariably eat glucose first, leaving lactose, a heavier molecule that requires more energy to break down, for dessert.

The question of how bacterial cells choose between glucose and lactose was the topic of the PhD dissertation of Jacques Monod, a charismatic French American biologist whose understanding of the inner workings of cells left an indelible mark on biology. Monod was studying at the Sorbonne, in Paris, in 1941, as World War II was raging across Europe. A committed communist and atheist, he turned his lab into a printing hub for Resistance propaganda. After the Allied forces landed at Normandy, he joined Charles de Gaulle's

French Forces of the Interior and took the fight to the Nazis, for which he received both the Croix de Guerre and the American Bronze Star.

Being obsessive and rational by temperament, after the war Monod chose to continue his investigations in his scientific post at the Institut Pasteur. In 1957, somewhat by accident, he was joined in this research by François Jacob, who before the war had aspired to be a surgeon but was stymied by the lingering effects of injuries. Less than a decade later, Monod and Jacob would share the Nobel Prize in Physiology or Medicine with André Lwoff for the fruits of their collaboration. Whereas Watson and Crick had discovered a structure for genetics, Jacob and Monod discovered the genes' logic of "chance and necessity."

To digest lactose, *E. coli* needs two pieces from the genomic catalogue in the form of proteins. One is lactose permease, a protein that sits on the cell membrane and, as its name suggests, draws lactose across into the cell. The other is ß-galactosidase, an enzyme that breaks down lactose into digestible units. From his earlier research, Monod knew these two proteins were not present in the cell when there was no lactose in the surrounding culture, but if lactose was added, the proteins would suddenly appear. He had also seen that if there was glucose in the culture too, the cell would always first gobble up the glucose.

When Jacob joined Monod at the Institut Pasteur, they set out to understand the habits, metabolic and otherwise, of cells, and to uncover them, they would hunt for *E. coli* with mutations that led them to make different choices. And mutants they found: some did not need lactose to make ß-galactosidase, others would not make the enzyme even in the presence of lactose, and some exhibited strange behavior, like starting to eat lactose even in the presence of glucose. Together these mutants configured the pieces of a puzzle that, when solved, led to a clear view of how *E. coli* made its choices.

Buried in the genome of *E. coli* are the genes—the instructions—for making the RNAs that are translated into lactose permease and ß-galactosidase, the proteins needed to process lactose. These stretches of DNA are not accessible to the cell's RNA-making machinery unless there's lactose inside the cell. This is because another protein binds tightly to the control region of these genes— the piece of DNA with information used to decide whether or not to make RNA—and prevents the DNA from being unzipped and transcribed. Not surprisingly, this protein is a type of transcription factor called a *repressor* because it represses the expression of information in a gene or genes.

When it is expressed, lactose permease sits in the cell membrane. It senses how much lactose is nearby, captures it, and then draws it inside the cell. Once inside the cell, the lactose binds to the repressor protein. This breaks the repressor's bond with those genes that code for the proteins to digest lactose and allows their expression. Permease is one example of how cells control genes based on their current context and environment. If there's glucose around, the cell emits a signal, in the form of proteins, that prevents the expression of ß-galactosidase even if the repressor falls off the DNA, acting something like a double-safe mechanism to ensure that glucose is used first. After the lactose is consumed, the repressor is back on the job, and the tools are put back in their toolbox until they're next needed. This complex system of controls is called the *lac* (lactose) *operon.*

Jacob and Monod envisioned the lac operon as a circuit, or switch, that bacteria can turn on or off to utilize lactose depending on each individual cell's own sensed environment. Later, other circuits in *E. coli* were discovered that are involved with other cellular mechanisms, such as consuming specific nutrients, fighting infections, or managing cell division, the function of each implying that it has a sensor hooked up to some sort of device for controlling gene expression. These are the simplest prototypes of how cells control the activity of genes.

Monod and Jacob argued that these circuits mediate specific functions and, as such, are universal features of cells: "Anything found to be true of *E. coli* must also be true of elephants," as Monod

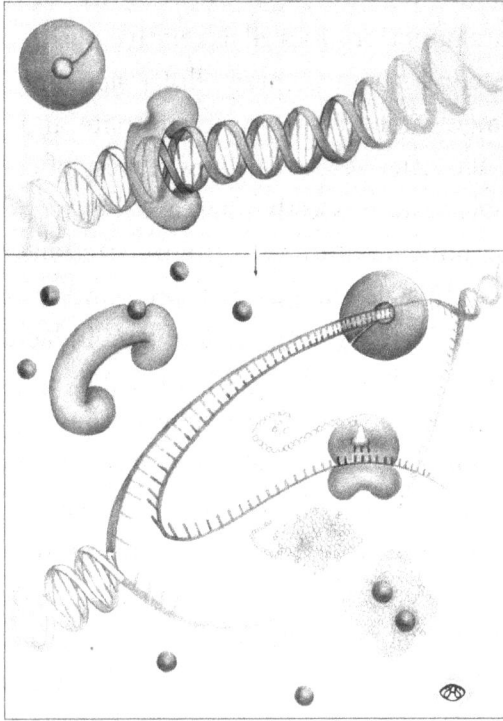

FIGURE 16. The working of the lac operon of *Escherichia coli.* The repressor (bean shape at the top) is bound to the DNA, preventing the process of transcription. The binding of lactose (small balls) releases it from the DNA and allows the process of transcription by an enzyme that synthesizes RNA for ß-galactosidase (large circular object). The RNA is translated into the ß-galactosidase protein (squarish shape at bottom right) that binds and digests the lactose.

put it plainly. Although an elephant is more complicated than bacteria, the logic that underlies their making must be the same, just as surely as a Concorde is more complicated than the Wright brothers' biplane, but their essence is the same.

LANDSCAPES OF FATE

To understand how Monod and Jacob's discovery applies to animals, it is important to revisit the key innovations that defined the transition to multicellular life on earth. First, there is clonal multicellularity, where an organism arises because the progeny of the zygote remains together with the cells diversifying, taking on specific roles and functions as well as relative positions in the space they create. Second, there is intercellular signaling, which allows cells to exchange information and coordinate and organize their identities and activities across this new society of cells. And third, there is a panoply of new tools that manage the creation of cell types and enable the implementation of the interactions between cells, which people in the field call *cell-type-specific transcription factors.*

Genetic circuits, such as the one involved in lactose utilization, are like automated robots used to help pick inventory swiftly and efficiently off the shelves of a sprawling warehouse. These robots' apps are programmed to know where to look for and move items so they can be packed up and sent to the customer who has placed an order, but the robots themselves aren't using the stuff. In a cell, the genetic circuits are a form of molecular robot or app that lets the cell, as the customer, receive delivery of the proteins it needs to maintain its shape, structure, and functions. The circuits are central to the cell's needs, identity, and workings, but they do not define what a cell looks like or what it does. They simply allow each cell to choose and use what it needs to build its architecture and to change that architecture depending on where it sits within a society of cells. As we have seen, each gene contains a control region that determines whether it will be expressed; the site where the repressor binds in the lac operon is one such site. These sites are like zip codes that order the activity of the genome. They are what the cell reaches into when it needs to express a gene and, in their own way, contain a signal, a switch to turn the circuit on or off.

The "everything store," Amazon, has a catalogue of more than twelve million products and sells upward of twelve billion individual items each year, so it's no surprise that it has two hundred thousand mobile robots working alongside one million people to help fulfill customer orders. *E. coli* has 4.5 million base pairs of DNA, about 6 megabytes of information, in its genome—the same magnitude of scale as Amazon's catalogue. So *E. coli* can configure a similar number of its own robots, in the form of genetic circuits, that the bacteria use to swim, eat, and evade predators. In contrast, we humans have 3 billion base pairs, about 2.7 gigabytes of information, which means our cells have the capacity to encode an enormous number of circuits. And like *E. coli*, human cells also use items in their genomic catalogue to make decisions about how and when to eat—though their choices are more complicated, because different cell types have different tastes and demands. This is represented in the control regions of the genes.

All *E. coli* cells are pretty much the same; our cells, like those of other animals, are different, each in its own way, and need large numbers of circuits to help them choose and use the genes they need to differentiate and function reliably. Muscle cells need a menu to help them build supple fibers and move; kidney cells need one that will support their filtering of toxic waste from the blood. In the brain, neurons consume three hundred calories per day to keep you thinking, eating, talking, and sleeping. Each type of cell assembles specific automated apps, made up of transcription factors, to fuel their specific functions. Apps are configured by shared control regions, and signals play a key role affecting choices and, importantly, coordinating the activity of the circuits across cell populations. They are integral to the emergence of multicellularity.

As is the case in other animals, many of the tools and materials available in the human genome are dedicated to building our bodies, allowing cells to create and organize their diversity under their own control. Recall how a change to a particular gene could create an organism with either a missing element or a part in the wrong place: a fruit fly with a leglike appendage in place of an antenna or a double set of wings. It is as if the robot picker goes to the right location in the genome, takes the part, and passes it along to the assembly line. If the wrong bit is chosen, it will be integrated in the wrong place and give rise to a faulty structure.

The generation of the numerous cell types of an animal during development is a slow process that follows a sequence of binary choices—*should I become A, or should I transform into B?*—not so different from the decision that *E. coli* makes between lactose and glucose.

Such choices draw upon many more genes than *β-galactosidase* and configure a cascade of decisions that build differences over time.

Let's envision a cell X that divides to generate two cells, each with the possibility of being either A or B. The one that chooses A will then have a second choice between becoming C and D. Meanwhile the one that chooses B will have to choose between E and F. A process of differentiation between A and B has begun, and at the end cell X's progeny will have produced a wide swathe of cells of various types, from skin to muscle to nerve. The essence of these choices is, as Monod would put it, the same as the choice to use lactose; each choice relies on a circuit that contributes to the overall genetic program. Genes lead to proteins, some of which are transcription factors, that search for control regions to activate other genes, some of which are transcription factors, and so forth. Control regions in A, B, C, D, E, and F determine which genes are activated depending on the sequence of transcription factors. This is how individual circuits become programs. Over the last few years, we have learned that every one of these decisions involves some-times hundreds of genes that endow each of those cell types with its characteristics.

At the same time, there are important differences from the workings of *E. coli*. These have to do with the large populations of cells that, in animals, are allocated in space and the need to coordi-nate these decisions between the many cells of a population across space and time. A and B are, more often than not, groups of cells. Yet, even considering the large number of genes involved in every decision, the underlying principle remains the same: cells receive and integrate signals, and then, using special proteins, they seek the control regions of genes that they need to switch on or off, as each activated circuit leads to another, and another.

British biologist C. H. (Conrad) Waddington memorably cap-tured the contours of this process, even before we knew all its details, by evoking an ordered journey through a mountainous landscape. At the far edge of the landscape, there is a summit, from which bi-furcating ravines and canyons roll down into deep valleys. An egg starts the journey to becoming an animal at this summit; the valleys represent paths to the final stages of cell differentiation, which lie in the deepest valleys at the bottom: the different cell types that comprise the adult animal.

At the summit, the egg awaits a signal—fertilization—that starts it rolling downhill at the same moment that it first divides and multiplies. At each fork, cells encounter a binary choice—left or right—still rolling downhill into one ravine or canyon after another, following a path to becoming neuron, muscle, or skin. Once a cell gets settled in the floor of a valley, it's nearly impossible for it to leave (unless, say, some scientists pull it out, as they do when they transfer a nucleus into an oocyte to generate a clone). Now, in order to shape this landscape of ravines and canyons, ridges must separate them. The depth of the grooves, the height of the ridges, and the flattening of the valley floors represent different features embedded in the processes of cell differentiation during development that will determine not only paths but also numbers of cells going in one direction or another.

As it starts its descent, the zygote divides into two cells, and these cells multiply into four cells, and so on. Not merely one but many cells face forks at the same time. Thus, the animal's cells progressively become distributed across different paths, in numbers that will allow the configuration of an organ and tissue. For example, a working heart requires defined numbers of cells in the atria and the ventricles, and alterations in these numbers can have dire consequences for the working of the heart. The steeper the descent into a ravine or canyon, the more likely a cell is to choose that groove, because it takes less energy to go with the momentum. The choice of one groove removes the opportunity to explore others, because the energy needed to climb up the landscape is prohibitively expensive. Partly this is because the cell uses transcription factors to close portions of the genome that it no longer needs to access after it's made each bifurcating choice, much as the repressor closes access to the *β-galactosidase* gene. This makes it more efficient in accessing needed tools without interference from those it does not need.

This allegory about cell differentiation and destiny, which biologists call "fate," is now known as the Waddington landscape. Waddington believed that the geography of the landscape had *something* to do with genes, and in one representation of his landscape, he placed genes as pegs and ropes underneath the hills and ravines, suggesting that their pulling and tensions determined the heights of the canyons and the ruggedness of the landscape. However, he also felt that whatever was going on to build organs and tissues, although related to gene activity, happened above the genes, and he used the term *epigenetic* to describe it. He said the landscape was an "epigenetic landscape." The term in his usage pertained more to development, though, as we have seen, it has been hijacked by a molecular view of events to refer to modifications of proteins that control the expression of genes. I prefer the classical definition because it refers to the activity of cells.

Then, in a way, yes, what is true for *E. coli* is true for the elephant. The apparent complexity of the animal cells and the way they come together in an embryo and an organism betray a relative molecular simplicity. There are more knobs, wires, and elements, but the principle is the same.

Following Waddington's lead, developmental biologists have invested profound effort into cataloguing the genes associated with every choice in the landscape—tying them to particular ravines and canyons and certainly to the decision to settle into a particular valley. Over the last twenty years, our understanding of genetics and our ability to manipulate genes have created a view of embryos and the development of an organism centered on the activities of genes and transcription factors. But Waddington was clear in his view that genes might not directly determine what cells do. Over the years the term *epigenetics* has evolved in meaning. It is now used to describe states of gene expression imposed by the environment of a cell or organism and recorded in the pattern of gene activity. Nevertheless, Waddington's insights about how cells make decisions persist.

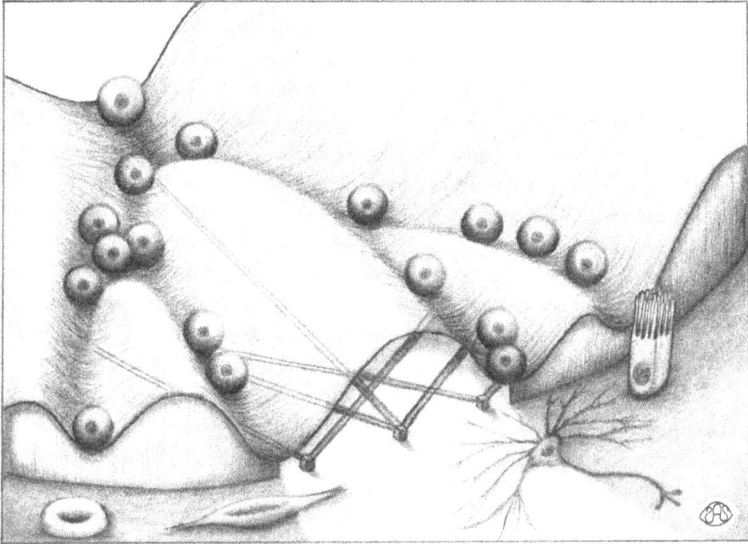

FIGURE 17. Representation of the process of how cells decide what to become during development as envisioned by Conrad Waddington in his 1957 book *The Strategy of Genes*. Cells are represented by balls rolling down a rugged landscape that is sculpted by the activity of genes, represented by pegs and ropes underneath. At the bottom, cells turn into definitive cell types, depending on the paths they have taken.

Waddington's landscape is also a representation of Monod's elephant. In its own oblique way, it captures the elements that underpin the transition from the genetic circuits of bacteria to the programs that underpin the development of animals and plants. However, as we have said above, an organism is no more a collection of genes than a house is a collection of bricks and stones. The epigenetic nature of the landscape clearly refers to the activities of the cells as the global output of the activity of the genome, and as there is order in the construction of a house or edifice, there is order and structure in the construction of an organism. We now turn our attention to this endeavor.

– five –

MOVING PATTERNS

WHERE DO BABIES COME FROM? IT'S A QUESTION THAT, sooner or later, every child ends up asking, though not every parent is prepared to answer straightforwardly. When I asked my own parents, they responded with a classic fantastical answer: a stork brings babies to their mothers' laps.

Oddly enough, many cultures share the tradition of bird-borne babies, stretching back to ancient Greece and Egypt, though each gives the tale its own spin. Sometimes the birds are herons or cranes rather than our customary storks. The location where babies are manufactured also differs depending on the culture: some trace the place of origin to marshes, others to gooseberry bushes, and still others to very specific sites, such as magical stones called *adebor-steine* that are found in the forest. My parents simply told me the birds carried them from Paris.

Viewed from a certain perspective, connecting our origins to birds is not so far-fetched, however. Before we even enter babyhood, we begin as embryos, and bird eggs have provided humans with a window into embryonic development for millennia. This window is not merely figurative: though bird eggs do not immediately present themselves as transparent, like the eggs of fish and frogs, you can carve a window in the hard exterior shell while keeping the embryo and its membranes intact. In this way you can watch what's happening inside as the egg is incubated. Aristotle did exactly this and drew from the experience to compose the first written account of the miraculous transformation of a small red speck near the yolk into a chick ready to hatch twenty-one days later. Aristotle had few concrete explanations for what he witnessed. He believed the red speck was a sperm activated by the hen's menstrual blood but did not know how the sperm became an embryo. He knew with certainty only what he observed: that organization progressively emerged where earlier there had been none.

Centuries after Aristotle, during the Enlightenment, scientists and philosophers were still seeking answers to the same questions. They discovered gravity, oxygen, and other elements, gave names to chemical processes, and created huge systems to classify the world around them. But the question of where babies come from remained strangely difficult to resolve.

An early Enlightenment thinker tackling the question of biological origins was the great English physician William Harvey, head physician to Kings James I and Charles I. From September to December, the period then known as the rutting season, King Charles went hunting for game nearly every week, providing Harvey the opportunity to hunt for eggs inside the does felled by the king and his retinue. He reasoned that if lizards, fish, frogs, and birds came from eggs, it stood to reason that deer and other mammals should as well. But the eggs were nowhere to be found. Throughout September and October, he found nothing in the wombs he dissected, even as dissections of does in November and December revealed the early outline of a developing deer, which was remarkably similar to the form visible through windows carved in chickens' eggs.

Harvey had located the embryo, though there was no sign of eggs to trace them back to. Still, in 1651, toward the end of his life, he collected his thoughts on animal development in a book titled *The Generation of Animals*. At the front of the book Harvey included an engraving of the Greek god Zeus opening an egg from which all sorts of animals escaped. On the egg was the motto *Ex ovo omnia*: From the egg, everything.

Harvey's insistence on the importance of the egg and his notions of how growth proceeded thereafter relied more on faith than fact, as he had not observed any direct evidence. Still, Harvey agreed with Aristotle: the chick's "parts are not fashioned simultaneously," he wrote, "but emerge in their due succession and order." Animal embryos were created through both the differentiation and the growth of the egg, and he called this process *epigenesis*—literally, *upon generation*, and the basis for the word used by Waddington for his landscape of fate.

A rival view held that a miniature version of the newborn organism, so small that it was invisible to the naked eye, existed from the start. Fertilization deposited this "preformed" organism near the egg's yolk, which provided the nourishment necessary for the organism to grow big enough to survive on its own. This notion, called *preformationism*, may sound irrational, but it gathered support after Antonie van Leeuwenhoek put semen under his novel microscope in 1677 and reported seeing independent organisms swimming around. Imaginations took off. Van Leeuwenhoek's animalcules were thought to each contain a miniature human being, which, during development, grew to life size. In one famous depiction, a baby is curled up in a fetal position within a sperm, its toes tipping toward the sperm's tail and its head comprising the sperm's head, waiting for the trigger to expand.

Taken to such extremes, preformationism's absurdity was inescapable. The miniature baby inside the sperm would itself have to carry sperm, which must then contain an even smaller baby. This required a lineage of microscopic matryoshka doll organisms, each one smaller than its parent, extending infinitely forward into future generations' gametes. Further, if all these miniatures were to grow to life size, in order to hold all generations, would not the egg's yolk need to become ever larger as the preformed organisms grew smaller? Or could the organisms ever become so small that they couldn't grow big enough to be born? This was also logically convenient for those more spiritually minded proponents, who, starting from the size of current sperm, performed an imaginary calculation of how many ever-larger organisms could be held in a sperm cell. In this manner, they were emboldened to trace humans back to the very first creation of every organism, at life size, in the Garden of Eden. In the end, it was impossible for something to come from nothing unless a divine hand had choreographed it. Preformationism assumed the first organisms had a divine origin, like Athena bursting fully formed from the head of Zeus.

An intellectual feud broke out between preformationists and epigeneticists, reaching its climax in the 1760s and 1770s. On the side of preformationism was Albrecht von Haller, a distinguished naturalist from Geneva known for characterizing the unique qualities of muscles and nerves. On the side of epigenetics was Caspar Friedrich Wolff, a young upstart scholar from Germany who had studied chicken embryos in painstaking detail for his doctoral dissertation in medicine. As part of this work, Wolff described the progressive emergence of the embryo's gut tube out of an undifferentiated gelatinous mass. He noted how matter inside the fertilized egg appeared to line up, form ridges and folds, and fuse together or cleave apart, creating not just the intestines but the blood vessels, heart, and other organs. He had "no doubt" that epigenesis occurred in chickens. He had seen it with his own eyes. "The intestine of the chicken embryo is first a simple membrane. This longitudinal stripe of the double plate, first plane, begins to swell into a cylindrical shape and then resembles the primitive intestine. We are thus certain, that this intestine is a new formation and that it could not have existed as such before."

Wolff dedicated his dissertation to Haller, which of course drew the elder scientist's attention. They disagreed fiercely but continued writing back and forth for nearly twenty years, until Haller's death, emphatically contesting the nature of development. Despite all evidence presented by Wolff, Haller insisted that every organ was preformed, explaining that existing microscopes were not powerful enough to reveal them.

Time would prove Wolff right, establishing his status as the founder of the modern field of embryology. But the dispute between preformationists and epigeneticists was not put to rest until the mid-1800s, when Rudolf Virchow and others identified cells as the essential, basic unit of all living systems. In the meantime, the study of animal embryos continued, producing some surprises of its own.

THE HOX STEP

In Königsberg in the 1820s, Estonian anatomist Karl Ernst von Baer filled his already cramped room with bottled specimens, surgical tools, candles, and microscopes. In his bottles, von Baer was preserving a collection of embryos from animals at each stage of development, carefully arranging them based on progressive changes in the form and structure of their organs, tissues, and overall bodies. He believed that identifying similarities and differences between species would eventually allow him to ascertain how animals are related to each other.

A moment of inattention delivered von Baer an epiphany he was not waiting for. One day he forgot to label two very young specimens. When he picked them up later, he couldn't figure out which was which; the embryos were almost identical. That's when it dawned on him:

> They might be lizards, small birds, or very young mammals. The formation of the heads and trunks in these animals is quite similar. The extremities are not yet present in these embryos but even if they were in the first stages of development, they would not indicate anything; since the feet of lizards and mammals, the wings and feet of birds, as well as the hands and feet of men develop from the same fundamental form.

Von Baer was looking at a bent, segmented structure with a clear sketch of a head and a tail. The head had the outline of two eyes, and the body had slits resembling the gills of a fish. Having gills was not enough to say it was a fish, because frog embryos would also have gills. But as he went through his collection to try to identify the specimens, he saw something arresting that, unbelievably, he had never noticed before. At the earliest stages of development, every embryo—from his fish and frogs to his cats and cows—had gill-like structures. In fact, they all looked eerily similar. From this insight, von Baer hypothesized that all animals develop from a common cast and only later in development are sculpted into the form of their specific species.

Giving order to nature was a preoccupation among naturalists in the nineteenth century. In 1830, two years after von Baer first reported the uncanny resemblance between early embryos, the French naturalists Étienne Geoffroy Saint-Hilaire and Georges Cuvier waged an intellectual battle of their own over the origin of the variety of forms seen across animals. Unlike the feud between Haller and Wolff, this fight was conducted largely in public in a series of debates before the Académie des Sciences in Paris. With a mix of wisdom and foolishness characteristic of the revolutionary period, Geoffroy proposed a "unity of composition" whereby animals, whether or not they were created by God, were all variations of a basic plan that could—and should—be uncovered. Cuvier, the highest authority on zoology at the time, disagreed, holding that every animal had been created independently by God, each one perfectly adapted to the environment and activities for which it had been formed. Geoffroy lost the debate by overreaching when he claimed that the bodies of mollusks such as squids and snails, which have no internal skeleton, possess the same organization as the bodies of vertebrates. This was too much for the scientific establishment to accept. Yet Geoffroy and his acolytes did not recant, contending that the wings of bats, paddles of porpoises, and legs of horses were simply variations of one grand design plan—as Charles Darwin wrote, with a nod to Geoffroy, in On the Origin of Species in 1859. In other words, although

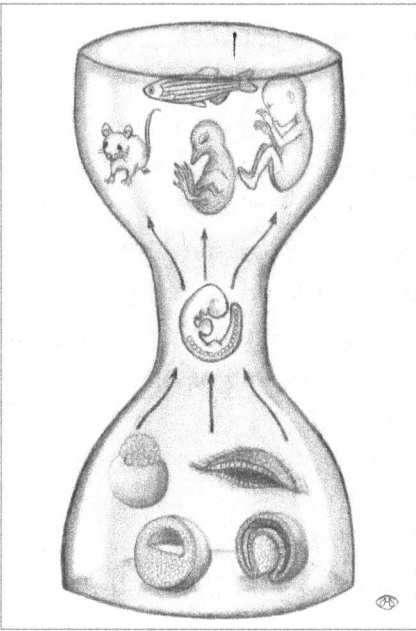

FIGURE 18. Denis Duboule's 1994 vision of the link between development and evolution as an hourglass. Following an initial idea of Karl Ernst von Baer, he suggested how, starting from very diverse forms and going through a bottleneck of similarity, animals diversify.

animals look wildly different at birth, they share commonalities, in both their basic body plan and most of their internal structures, that can be seen in their embryonic development.

In 1994, Denis Duboule, professor of developmental genetics and genomics at the University of Geneva in Switzerland, introduced the notion of a "developmental hourglass" to describe the progression from eggs, in all of their variation among species, to the similarity of early embryos. Duboule noted something curious: von Baer's epiphany mapped to the neck of the hourglass, and this moment corresponds with the time in which *Hox* genes—those tools used to distinguish different regions of the body axis of all animals—are fully expressed along the head–tail axis in bilateral animals like fish and frogs and us. On either side of the hourglass's neck, animal cells can explore organizing themselves into many different shapes and forms, but at the neck, Duboule suggested, they are bound to look the same, because they must use the tools available in the *Hox* genes as they start to build the animal, and these tools, for some mysterious reason, are absolutely conserved.

As metaphors go, this is a good one, but it raises the question of what exactly about the *Hox* genes causes a convergence in shape, eliminating the differences between chicken, frog, fish, and human eggs. How do objects that start out differently become the same as they multiply and grow? The answer, I believe, lies in the dances cells engage in early in development. Rather than as a rite of passage, we might think of this as the cells' "Hox step"—a common dance step, like the box step, which is a building block of everything from the waltz to the rumba. But to see the choreography in action, we need to return to the very beginning of the dance, the creation of an animal's first cell, the zygote.

CELLS AT THE BALL

The entry of a sperm into an egg triggers a cycle of cell division and multiplication that rapidly generates thousands of cells. In all animals, this first act of development ends with a mass of cells indistinguishable from each other—called the *blastula* or the *blastoderm*, from the Greek roots meaning *sprouting skin*—which are the source of everything that comes after.

At this stage, the cells huddle tightly together in an epithelium, the wall-like shape that cells sometimes adopt, determined by the structure of the original egg and its yolk, which of course is the cells' primary food source. Typically, the cells either sit on top of the yolk of the egg, forming a flat disc, like a raft floating on a sea of food that they can dip into, or engulf the yolk, forming a sphere around it and consuming it from the outside in, like mold on an orange. In both formations, the cells are at first sight all the same. They don't know what they are going to do or become, and so they remain unspecialized.

Then, at a time precisely measured for every species from the moment of the egg's fertilization, some epithelial cells transform themselves into a more supple form, a *mesenchyme*, and start to move around in a manner reminiscent of a dance. They begin by moving in small clusters, as if obeying the baton of a hidden conductor, implementing a set of movements that will transform the undifferentiated mass into the first rough outline of an organism. As the dance proceeds, the cells divide into groups as more and more cells from the epithelium loosen up and join in. Moving away from the palisade, some head toward the interior of the egg; others jostle for position on the surface, moving in ways that bend the original palisade, much of which still exists, to create another wall of cells that will bar the movement of some of their fellow cells. By the end of these movements, von Baer's uncanny structure will be visible—an embryo with a headlike form at one end, a tail at the other, and groups of cells, or segments, running between and within the layers that are starting the process of differentiation into various tissues and organs. This structure is always alike for every animal within each major phylum. We are now at the neck of Duboule's hourglass, the *Hox* genes have been unfolded, and the cells have performed their "Hox trot."

The technical term for the cellular choreography that brings us to this moment is *gastrulation*, derived from the Greek root *gaster*, meaning *stomach*, which gives us the word *gastronomy*. It's an awkward word coined in 1872 by Ernst Haeckel, a professor at the University of Jena in Germany. When Haeckel was studying the development of sponge embryos, which start out in the form of a ball, he noted that a group of cells moved to create the lining of the gut by folding in on the ball. Imagine a slightly deflated balloon into which you can push your finger to create an indentation with a second palisade of cells. That indentation into the balloon's sphere is the sponge embryo's so-called primitive gut, or *gastrula*, as Haeckel named it, and for this reason the process is called *gastrulation*.

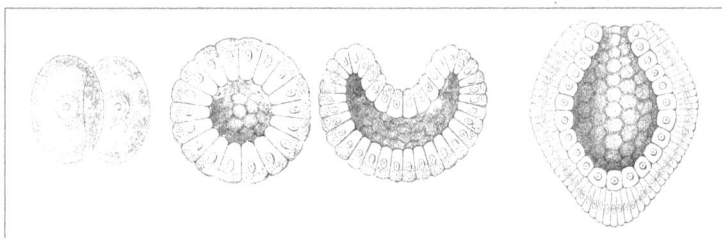

FIGURE 19. Ernst Haeckel's drawings of gastrulation in sponges. The process proceeds in stages illustrated here from left to right. From "Studien zur Gastraea-theorie 1877," 1879.

Based on his observations, Haeckel proposed that gastrulation happens early in the development of all animals. He was correct, but sponges were not a good representation of how it plays out in most animals. In sponges the cells divide into only two groups, or layers, during gastrulation. In most other animals, they divide into three. These are called *germ layers* to indicate that they are the seed-bed for what will come. The structures that will ultimately emerge from each germ layer begin to be distinguished by the appearance of their cells, their movements, and the positions they take at the end of their dance relative to each other. The gut and associated organs, including the liver, pancreas, and lungs, arise at the end of gastrulation from the *endoderm*, the innermost layer. The skin and the brain arise from the *ectoderm*, with a final position as the outermost layer. Between these two layers lies the *mesoderm*, which will give rise to the muscles, including the heart, blood, kidneys, and genitals.

In real time, the dance of gastrulation appears to consist of almost imperceptible changes and movements. This is why it appeared to Wolff and others that the chicken embryo materialized, as if from thin air, from amorphous matter. But today, we can film the cells' behavior and speed it up to reveal the magic of the appearance of the organism, tracing groups of cells shapeshifting and Hox-stepping together. We see cells moving sometimes alone, but often in groups, displaying purpose and precision in their movement. In chicken embryos the initial dance resembles the polonaise, where pairs of dancers promenade in mirrored circles up and down the ballroom. The line between the two sides of dancers reflects the bilateral body plan, across which we can see the head and tail, back and belly.

In insects, the rules of the dance are inscribed in the egg before it is even fertilized, with all the directions for building the body plan laid out from the creation of the very first cell. The fruit fly egg, for example, contains landmarks in the form of specific proteins precisely distributed within its cell space, signposting what will be the front and back, top and bottom of the embryo. After fertilization, as the cells emerge and position themselves across the egg, they follow instructions from these landmarks about what they should become. Twenty-four hours later, after executing specific genetic programs, the cells have built a maggot that crawls out of the egg. In contrast, there are no signposts in vertebrates. Here the cells create the rules of their dance together, out of nothing; it looks like magic, but it is emergence. Cells exchange signals to create the body's axes as a reference for the organization of the cells, set the tempo of the music, and dole out roles to each other based on their relative position following an invisible compass that leads them to and fro. When the dance of gastrulation stops, it reveals the outline of a frog, a rabbit, a pig, or a human. In all cases gastrulation is about cells, about their moving around with purpose and direction relative to each other, sensing their numbers and the geometry of the space they build. They use their cytoskeleton to create forces, to move, and chemical signals to tell each other what they are doing, what they should do, and their future fates. They come together in groups that go solid to fluid and back to solid, with a surprising artistic sense that can only be expressed as a wondrous combination of dance and sculpture. Transcription, the process of gene expression, plays some role in this but always at the request of the groups of cells. The cells show their building ability in gastrulation more than in any other process.

Through gastrulation cells sketch out organisms, and this is why my charismatic and insightful colleague, the late South African engineer turned biologist Lewis Wolpert, said, "It is not birth, marriage or death, but gastrulation which is truly the most important time in your life."

In each species this choreography has a distinct physical reference point, a place within the mass of cells where the dance begins. Regardless of where exactly that point lies, it will ultimately become the organism's back end. In frogs it appears first as a slit, called the *dorsal lip*, that develops into a hole in an otherwise indistinguishable ball of cells. The endo- and mesodermal cells use this hole as a reference point to ingress, gently and deliberately burrowing into their locations from head to tail. Fish undergo gastrulation in a similar manner, but something happened when animals moved

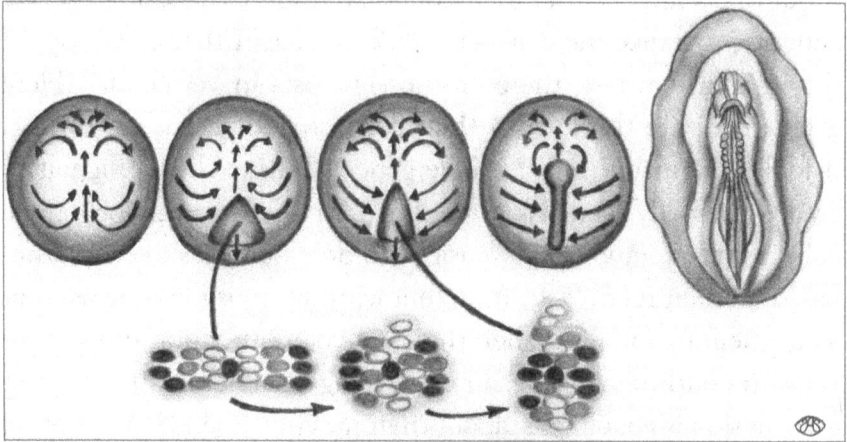

FIGURE 20. Gastrulation in the chicken embryo. Initially the embryo is a large disc of cells that engage in coordinated movements that, globally, resemble a polonaise dance (*arrows*), drive the emergence of the primitive streak (*triangle at the bottom*), and after stereotyped movements and rearrangements, give rise to the main body axis of the embryo with the head at the top and the tail at the bottom (*extreme right*).

to land. The hole became a groove, and the cells changed their behaviors. In birds, whose initial mass of cells form a disc rather than a ball, a dimple emerges at one edge of the disc, where cells begin to plough a furrow toward the opposite pole of the disc. This feature is called the *primitive streak*, and like the dorsal lip in frogs, it is used by the endo- and mesodermal cells to colonize the emerging embryo with layers and groups of cells that will become organs and tissues. Mammals also have a primitive streak that, as we shall see, has become a landmark in defining what makes a human being. Once all the cells have taken their final positions, the spot where the dance kicked off becomes the anus, the hindmost end of the gut. This special hole is thus a souvenir of the most important moment in your coming into being.

I've mentioned that prior to the dance, the cells are indistinguishable from each other. Yet, as soon as the dance commences, the cells in each of the three germ layers move in different ways, reflecting their fates even before we can tell what they will be. Those that will become the endoderm loosen up and burrow deep into the emerging embryo, where they form a tightly packed sheet, roll up into a tube, and seal themselves into the gut. Those that will become the ectoderm remain outside the disc or ball of cells, waiting for the moment to begin to build the nervous system.

Most remarkably, the mesodermal cells lose their wall-like appearance, disconnecting from each other and moving with a clearly individualistic sense of direction to spread themselves betwixt and between the endoderm and the ectoderm. Mesodermal cells exhibit excellent navigational skills and sometimes even flocking behavior, depending on what part of the body they will eventually become. In mammals, the mesodermal cells for the heart move to the most forward region of the embryo, and those that will give rise to the surrounding structure of the chest spread out behind them, then form tight balls called *somites* along the front–back axis of the embryo. The somites are akin to buds for the muscles and ribs.

An intriguing feature of gastrulation is explicit in Duboule's hourglass view of development—namely, that convergence of design at the neck. Every embryo starts from a different egg, with a different geometry and different physical constraints, and will execute gastrulation in a different way, but the outcome of the process is so similar that embryos of different organisms look very similar. This stage, first described by von Baer, contains some attractor of form that we need to understand.

While the tempo and timing of this complex choreography are specific to each species, within a species it is exactly the same from one embryo to another. If you lined up a hundred frog or chicken embryos to watch them develop side by side, you'd witness synchronous movements in every one of them (unless something went wrong in the development of a particular organism). It is a wondrous sight. You can even stain groups of cells from different parts of the initial mass of cells in order to follow them through the entire process of gastrulation. Thanks to these experiments, first performed on frog embryos, we have learned that the position of a cell before gastrulation determines where it ends up in the embryo: cells in the same position end up in the same place. Every cell's fate can be mapped based on where it came into existence.

One special group of cells across all animals is deaf to the music of gastrulation: the germ cells, which will later give rise to sperm or eggs. It is at this moment, before cells begin to roll down the Waddington landscape, that the Faustian pact between eukaryotic cells and the genome is made official. During gastrulation, the germ cells are set aside to protect an unused copy of the cellular toolbox for use by the next generation. One property of germ cells in all animals is their ability to find their way into the gonad, when it is made. Until then they are tucked aside, impervious to all the dancing going on around them, and when the time arises, they engage in a miraculous journey through the embryo until reaching the gonad, where they will become the gametes to move into the next generation.

THE LANGUAGE OF CELLS

In a waltz, foxtrot, or rumba, the dancers' movements follow certain rules, but the dancers have a lot of discretion in how exactly they execute each step. In a ballet, the dancers' choreography is precisely scripted, with the direction and pace of movements dictated in advance to ensure all the dancers are coordinated throughout the performance. The complex movements of gastrulation are more like the Imperial Ballet's *Swan Lake* than a twirl around a ballroom to "The Blue Danube," repeating themselves from embryo to embryo without missing a step. There must be some sort of composition, with clearly marked directions for movements to be taken at specific measures and bars aligned with the musical score. There may even be a choreographer, though not a divine one.

The precision timing and movement of gastrulation was first confirmed in newts in the Berlin lab of Hans Spemann. Between the world wars at the beginning of the twentieth century, there was no better place to seek out answers about the deepest nature of embryos than Spemann's lab. He conducted experiments using newts because they were easy to obtain; plus, their large, yolky eggs develop quickly into a mass of cells that can be observed at low magnification and are easily manipulated. And manipulate them he did—he reportedly said that he spent the bulk of his life placing embryos in ever more embarrassing positions, removing or mixing cells and seeing how the cells resolved the set awkwardness. This forced the cells to have conversations about their fates on a schedule that he dictated rather than following their usual choreography. From these experiments, we learned the patterns of cell movements and how they relate to cells' fates.

After working along these lines for several years, he turned his attention to the dorsal lip, a slit that acts as a reference point for the cells' movements during gastrulation. In 1921, he asked Hilde Mangold, a brilliant student in his lab, to repeat an experiment he'd done a few times already with a curious outcome: when, just before the start of gastrulation, he removed tissue from around the dorsal lip in one embryo and grafted it into another embryo, in a place opposite to its own dorsal lip, a second, fully organized body emerged—often as a conjoined twin. No other group of cells that he'd tried had the same effect. It might be that the new body was produced by the grafted cells, which "remembered" what they were supposed to do and went ahead with their upcoming steps, despite having been relocated into a different spot in a different embryo, but there were other possibilities as well.

To test this possible explanation, Mangold chose to repeat the experiment using cells from two species of newts whose cells had different pigmentation. The color of the cells allowed her to easily see which cells came from the donor and which from the recipient when the twins developed. Amazingly, the twin embryo was growing from the recipient's cells. The transplanted dorsal lip cells changed the identity of the recipient embryo's cells, *instructing* them to move and change to give rise to parts of the body they normally would not. The tissue around the dorsal lip is now known as the *Spemann-Mangold organizer*, as it provides instructions for how the cells around it should organize themselves. Tragically, Hilde Mangold died in an accident and was not awarded, with Spemann, the 1935 Nobel Prize in Physiology or Medicine for making the discovery.

Cells with a similar ability to choreograph the movements of other cells during gastrulation were later discovered in the embryos of birds, rabbits, and mice, though the process in humans has mostly remained a mystery due to reasonable laws against experimentation on human embryos. It is possible that conjoined twins result from malfunction or splitting of the organizer. However, even lacking precise information about human gastrulation, researchers have been able to discern that the language used to build the blueprint of the organism is universal and that embryos can even translate signals across species lines. If you take the Spemann-Mangold organizer from a frog embryo and place it in a chicken embryo in the early days of development, it will instruct the surrounding cells to construct a second embryo—and the embryo of a chicken, not a frog, will develop. The same is true if you cross the organizer of rabbits and chickens. The organizer sends an instruction that can be read and interpreted by the recipient cells in their own language. This is very reminiscent of what we have seen with the PAX6 gene, which, independently of its source, will make the eyes

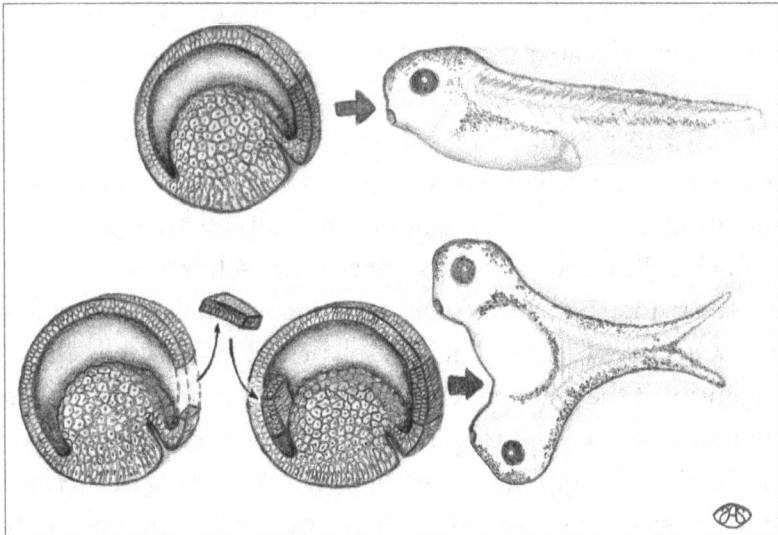

FIGURE 21. The Spemann-Mangold experiment, performed by Hilde Mangold, in which transplantation of the organizer induces the host cells to generate a secondary embryo.

of the organism in which it is expressed. It is the cell that reads, interprets, and translates the tools or signals it is given.

Similar conversations between cells take place while organs are emerging. Some groups of cells instruct and lead; others listen and enact. Take, for example, the case of hen's teeth. Like many organs, teeth start developing in animals as part of a conversation—what biologists like to call an *interaction*—between two types of tissues.

In the developing jaw of a chicken, the interaction is between a loose ensemble of cells, a *mesenchyme*, which can give rise to bones, muscle, or blood cells, and an adjacent cluster of epithelial cells that have developed from the ectoderm. If you substitute mesenchyme cells in the developing jaw of a chicken embryo with mesenchyme cells from a mouse, teeth emerge.[1] Most surprisingly, the teeth that arise are reptilian, very much like those of a crocodile, a real Jurassic Park sprung up in the lab.

If we return to the cell's-eye view of life, in which the genome is a catalogue of hardware available to cells for building and managing an organism, the mystery of hens' reptilian teeth becomes less unusual. The fossil record and genetic studies have revealed that birds are living dinosaurs that lost their teeth more than eighty million years ago. Their fossilized ancestors include the likes of the birdlike dinosaur *Archeopterix*, which had jaws studded with sharp pin teeth. If there were genes for teeth, Richard Dawkins's theory of the selfish gene dictates that those genes would have been culled from the chicken genome a long time ago. In evolution, what you don't use, you tend to lose. Yet the genes for teeth remain, untapped but available, a set of tools in storage but imbued with multiple potential uses. Birds are only missing some biochemical incantation to bring this hardware together to regrow their teeth.

In fact, a closer examination of the cases in which birds are naturally born with teeth offers clues to exactly what is going on. Chickens with a mutation called *Talpid* exhibit many odd features, including a deformed head and transparent, reptilian teeth. Geneticists have found that *Talpid* mutants have high levels of a signaling protein called Sonic Hedgehog (or Sonic, for short). Experiments have shown that Sonic signals are a key channel through which epithelial and mesenchymal cells in the region of the jaw communicate with each other to build teeth. In most birds, Sonic is silent in these cells, but in *Talpid* mutants it's present—and chattering "Sonic, Sonic" loud and clear.[2] Providing this protein to the epithelial cells while the jaws are forming reignites a conversation that was active millions of years ago but has since gone silent. This is exactly what the mouse mesenchyme provides.

The magic of gastrulation also involves a chorus of conversations between cells. Cells exchange huge amounts of information, changing their movements and their dance partners again and again in response. The exchanges are so swift and voluminous that they might easily descend into cacophony. But they don't. Instead, the lead is taken by three signaling proteins—BMP, Nodal, and Wnt—that coordinate the exchanges to the final step of the dance, when the cells are lined up in a body plan that looks alike in all vertebrates. These three proteins are involved in deciding where the head and tail, front and back of the organism will be. Their initial role is to activate the *Brachyury* gene, which has the code for a transcription factor that unlocks thousands of genes in the cells' catalogue of tools and materials. When *Brachyury* is unlocked, cells gain access to a variety of hardware involved in movement and orientation and in the maintenance of the signals that keep them going, Nodal and Wnt. These proteins are at the heart of the structure and activity of the dorsal lip in frogs; they sculpt the primitive streak in chicken and mice and plough the front–back axis of the embryo, moving and swapping dance partners to explore where and what they will become, feeling out new combinations, and trying out new levels of signaling, sensation, and response. Throughout gastrulation and in an orderly manner, cells reach into the genome for the genes that, when transformed into proteins, sculpt the outline of the organism allocating cells to specific places and fates.

The acronyms and enigmatic names of the genes and proteins used by cells to build the body plan hide their role in cells' signaling language and reflect the way they were discovered. The name of the

gene *Sonic Hedgehog*, referred to as *Shh*, is the vertebrate analogue—
or *homologue*, as geneticists refer to them—of the gene *hedgehog* found
in *Drosophila*, where the body of the mutant fly maggot resembles a
miniature hedgehog. *BMP* stands for Bone Morphogenetic Protein,
so named for its role in the growth of bone, where it was first discovered. *Nodal* is produced in early embryos at the node, the equivalent
of the organizer in birds and mammals, where it distinguishes left
from right. And *Wnt* is an acronym created after researchers realized
that two previously identified genes were actually the same—the
one in *Drosophila* called *wingless*, because the original mutant lacked
wings, and a tumor-promoting gene called *Int1*, which causes cancer.

The mechanism that guides the crystallization of these patterns
is only starting to become clear; it appears that the ability of the
cells to read and interpret forces generated by numbers and geometry, together with these signals, allows them to access what they
need from the genome to build the embryo. Watching this process,
it's hard not to have the impression that cells always know what to
do and where to move relative to each other, to find their positions
in space so that the cells of organs, like the heart and lungs, work
together and their plumbing fits.

Positioning relative to other cells turns out to be one of cells'
most valuable assets and is very much in evidence in the actions of
a wondrous group that arises shortly after gastrulation: the neural
crest. These cells give rise to large swaths of your body. Neural crest
cells make up the bone and cartilage of your head and contribute to
your heart, your teeth, parts of your gut, your eyes, and your muscles.
They are the source of the pigment cells that give hue to your skin
and, in other animals, create the intricate patterns of stripes and
spots that enthrall us. These cells are born in the most dorsal part
of the developing nervous system under the influence of Wnt signaling, spanning the length of the body. They then migrate out to
distribute themselves throughout the body, to become the different
kinds of cells depending on where they land. For example, the patterns that catch your eye when looking at a zebra or a tiger reflect
the paths taken by neural crest cells that give rise to pigment cells.

The migration of these cells is as precise as it is mysterious, following cues that must be hidden in the cellular territories they traverse. Recent studies suggest that these cues are not simply chemical; instead, inputs include the hardness or softness of the territories and the density or looseness of the cell populations they invade. Neighbors are crucial in this case and influence what the cells become as these mechanical properties are translated into gene activity, thus expanding the language of the cells.

COUNTING FINGERS AND TOES

Most of us have five fingers on each of our hands—a thumb, an index finger, a middle finger, a ring finger, and a pinkie—and a similar pattern of toes. Some have six, a condition that is not so unusual, called *polydactyly*, meaning *many fingers*, which, as we have seen, was the first human trait determined to be inherited. The pattern and order of these digits are decided very early in development, shortly after the formation of the buds on either side of the trunk, which give rise to limbs and which we have in common with other vertebrates, and long before there's a hint of a toe or finger. Furthermore, although your hands are a different size than mine, your fingers and my fingers are proportional to the size of each of our hands; their patterns scale to the size of the limb. These simple observations haunted Lewis Wolpert of the celebrated quip about gastrulation. Mulling over how patterns like this come to be, in 1969 he landed on the idea that cells either receive or enact instructions about what they do based on their position within a group of cells. He called this *positional information*.[3]

To tease out the implications of position and scaling, Wolpert conducted a thought experiment. He considered what other familiar objects also have the property of always maintaining consistent proportions in their patterns, regardless of their overall size. The answer he found was a national flag, and he latched on to the French tricolor; actually any tricolor flag will do, but he chose the French one. Whether the flag is large or small, the blue, white, and red rectangles always occur in the same relative position, with the blue rectangle closest to the mast and the red one furthest away. The rectangles are always equal thirds of the whole.

Wolpert then turned his attention to a real-life situation featuring a pattern of three elements ordered in space. The digits in our hands and feet were a good choice, though they comprise five elements rather than three, but the wings and feet of birds fit the analogy better and would allow him to do experiments. Chickens, for example, have three claws—the equivalent of an index finger (digit 2), a middle finger (digit 3), and a ring finger (digit 4). His choice was inspired: the digits of a developing chicken embryo were like the rectangles of the French flag—digit 2 was the blue field, digit 3 was the white, and digit 4 was the red. In a tricolor "flag" made up of cells, something was signaling to groupings of cells whether to be blue, white, or red—digit 2, 3, or 4—and how large to grow in relation to the other groups of cells in the whole.

Continuing with this thought exercise, Wolpert imagined a blank sheet made up of cells and suggested that the instructions for the pattern to be built came from some mysterious chemical instruction. He called this unknown substance a *morphogen*, or generator of form. This morphogen would leak from one side of the sheet and diffuse across it, generating a gradient. Then—and here was his great insight—he posited that different concentrations of the morphogen would be read by the cells, with high levels interpreted by the cells as a signal to be blue (or 2), intermediate levels as a signal to be white (or 3), and low levels as a signal to be red (or 4). This is another type of cellular conversation, where the meaning of the message to a cell depends on how far away the cell is from the source—a bit like reading a smoke signal as warning of a less immediate threat if it's a small wisp in the far distance versus a billowing plume just over the next hill. Wolpert assumed that the morphogen's message was independent of the size of the "flag," meaning that people's pinkie and index fingers generally have the same relative proportions, whether they've got small hands or large ones.

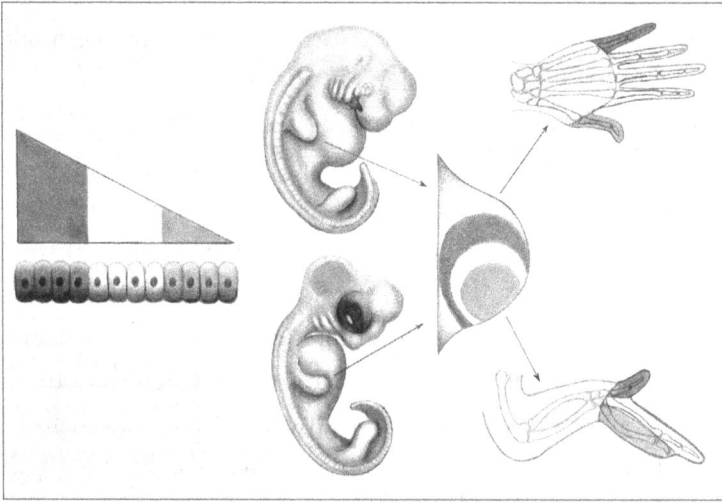

FIGURE 22. The French flag model of positional information, according to which a gradient of a substance, a morphogen, instructs cells as to their different fates. The vertebrate limb bud has provided a useful example of how this works in embryos. A substance diffuses from the posterior region of the bud, and by reading concentrations, cells are instructed to give rise to the different digits. Even though the morphogen is the same, its readout is different in humans and birds.

Multiple experiments have been done to test Wolpert's French flag model of cell signaling. As in Spemann-Mangold experiments to identify the organizer role of the dorsal lip, cells were moved from an embryo and grafted into a recipient to see what they would do, but this time the group of cells came from the limb bud. Tucked away at the back of each bud are some cells with a remarkable property: if you take these cells from a chicken embryo and place them in the front part of a normal limb bud in another chicken embryo, the resulting chick will have a wing with six digits instead of three. Moreover, the extra three digits are a mirror image of the normal ones, so the wing has a pattern of digits, from front to back, of 432234 rather than the normal 234. The group of cells with these special powers has been named the *zone of polarizing activity* and represents a sort of Spemann-Mangold organizer for the digits. When it came time to look for a molecule that diffused across that group of cells, one popped up: the protein Sonic Hedgehog.

Recall that *Talpid* mutants have teeth because they have too much Sonic, a cacophony of loud, disorganized signaling that the epithelial cells in the jaw can't ignore. In these mutants, many other tissues, including the limb buds, also have too much Sonic. These mutants were named *Talpid* after the animal family Talpidae (which includes moles), which have an extra digit, the *os falciforme*, or false bone, in addition to their normal fingers and toes. Further studies have shown that if you place a bead laced with Sonic at the front end of a chicken embryo limb bud, the chick will be born with 432234 polydactyly. In humans, many cases of polydactyly are similarly associated with having Sonic in the wrong place—a scrambled conversation, as in the game of telephone, or a misplaced signal, as in the Boy Who Cried Wolf.

The syndromes associated with excess Sonic make clear that different tissues interpret the same signal differently and that the function of the signal is not so much to instruct but to *organize*, to define the domains in which cells exercise their options. Whether Sonic comes from a mouse, fish, or bead, if it's placed in a chicken limb bud, it will inspire cells to build extra digits, and if it's placed in the mouth, it will inspire them to build teeth that the organism's ancestors haven't had for millions of years.

How cells decode and translate these signals is a complex story in its own right. It is not, as Wolpert first envisioned, a simple reading of local concentrations, like reading a gas or water meter. Rather, it's an intricate operation that depends on what's already happening inside the cell and how the proteins are linked and work together. From the moment of gastrulation, cells in different locations *are* different because of who their neighbors are and the conversational partners they're encountering, and these uniquely individual experiences lead them to express specific genes and thus look different.

These differences can be read in cells' molecular makeup. Within each cell and each type of cell, different molecular machines, proteins, and lipids are themselves reaching out for tools and fix-tures. The conversation is happening at many levels, and the same signal will have a different effect on different audiences—the same way that shaking your head side to side means yes in India and no in Europe or a thumbs-up conveys encouragement in the United States but is highly offensive in many areas of the Middle East. In the same way, Sonic means teeth in the cells of the mouth and digits in the end cells of limb buds.

Still, Wolpert's concept of positional information provides a way to think about how groups of cells are able to create consistent, matching patterns across the vast distances of a body. Look at your own body, and you'll observe that your organs and tissues have pat-terns with respect to their particular position in the body, as well as being internally patterned. Your eyes, nose, mouth, and ears are clearly positioned relative to each other, despite being on opposite sides of your body. Their cellular seeds were laid out following the rules of positional information. In addition to Sonic, there are BMP, Nodal, FGF, Wnt, and a few more, all used to reach out into the toolbox that lies in the DNA and build different cells that then shape themselves into organs and tissues. In fact, it is the repetitive use of these signals in different processes that tells us that they don't instruct cells; rather, they steer along their programs of gene expres-sion. Together with geometry and mechanics, these signals decide where specific groups of cells go, the shapes of organs, and the num-bers and proportions of cells that make them. The experiments tell us this, but we still don't know much about how cells do it.

It's true that the signals from Sonic, BMP, Nodal, Wnt, FGF, TGF, and others come in the form of proteins whose code resides in genes, but the genes do not provide the cells with any instructions about why that signal means they should move to a certain place, build tissues of a certain size, or take on a certain shape. The genes are agnostic about anything except the protein that will be made after they're copied into RNA, and the genes that are copied are copied because of signals being communicated between cells based on their environment and exchanges with each other. Genes are the instruments of the cell, a literal alphabet from which they construct their chemical language. They have learned to "count" by measuring the strength of chemical signals.

TIME, SPACE, AND PLUMBING

As we can glean from what we have learned about positional information, everything is relative in the world of cells. How time and space are partitioned and controlled matters to the development of organisms.

In animals—and, in particular, in vertebrates—the dance of gastrulation comes to an end once the basic body plan of the organism has been sketched out, with the rudiments of the main organs and tissues all in place, ready to be sculpted into their final shape. For example, at this time the endoderm covers the length of the embryo's front side, with cells in position to fold into a tube that will later be subdivided into the esophagus, lungs, pancreas, liver, stomach, and intestine. In each of these territories, positional information and signaling help the cells to establish a pattern in their individual structural type and their connections to create functional entities.

At the end of gastrulation, the seeds of the components of the body are in place, but in birds and mammals, like mice or us, the body is foreshortened and ends somewhere at the level of the forelimb. Between that point and the anus there is nothing—yet. The missing space is created by a growing mass of cells located at the posterior tip that is packaged into a structure, like a thick sausage, elongating from anterior to posterior. This structure sows the spinal cord and different derivatives of the mesoderm that will give rise to the limbs, the kidneys, and the gonads at different positions of the anteroposterior axis. These positions are determined by our old friends the *Hox* genes, whose expression unfolds as the body grows.

On either side of the spinal cord one can see the progressive appearance of pairs of cysts of mesodermal cells called *somites*, which serve as a kind of yardstick for the growing body. These cysts will

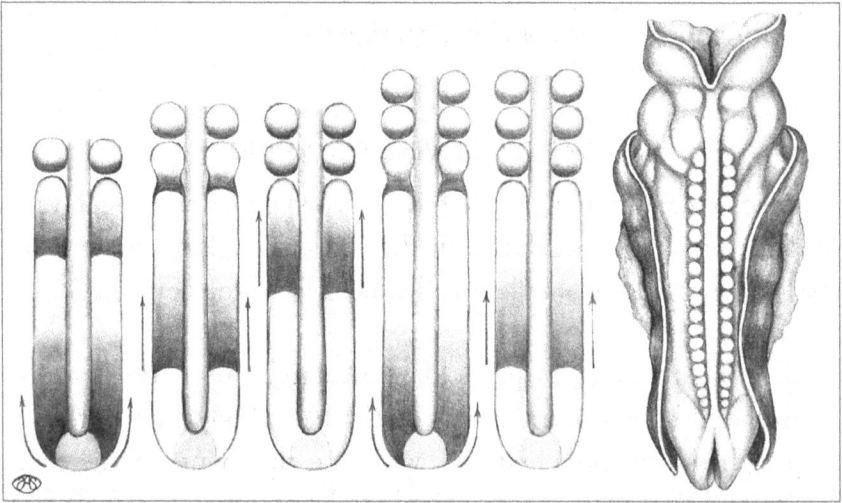

FIGURE 23. During the process of somitogenesis, waves of gene expression proceed from posterior to anterior (indicated by arrows) to hit a level of the body at which oscillations stop and cells form a cyst, the somite. These emerge sequentially following the rhythm of the waves, leading to a series of segments that will be sculpted into the ribs and muscles of the body axis. *Right:* an approximately five-week-old human embryo with somites.

later give rise to the muscles and ribs of the trunk as well as to the vertebral column. For the present, they arise rhythmically, in pairs, one after another, with a timing that is characteristic of every species: thirty minutes in the zebrafish, three hours in the mouse, five hours in a human. The process is called *somitogenesis* and reflects the extension and growth of the body. By the time the process ends, the body is complete: at sixty somites in a mouse, forty-two to forty-four in a human, five hundred in a snake. Yet the embryos still look superficially very much alike. A series of somites bearing just such a resemblance caught Karl Ernst von Baer's eye when he was trying to figure out which embryos his unlabeled specimen bottles contained.

The rhythmic appearance of the somites prompted the suggestion that they might be the result of some repetitive process, perhaps a periodic wave, that sweeps through the cells from posterior to anterior across the growing mesoderm and allows it to grow from anterior to posterior. This thought became a fact in 1997 when developmental biologist Olivier Pourquié, at the time working in Marseille, reported periodic activity of a small number of genes, across the elongating body axis of chicken embryos, perfectly in tune with the process of somitogenesis. The expression of the genes, associated with Notch signaling, painted waves of gene expression within the growing mass of mesodermal cells that fuel the appearance of somites. The waves repeated themselves periodically and stopped abruptly at the place where the cysts formed, like waves breaking on a beach and leaving their salty impression in the sand as in a fast-receding tide, their periods and rhythms in step with the appearance of somites. When the wave hit the front, it stopped, and the process started again from the posterior end, converting time into space. Of course, this is happening against the background of a constant increase in the number of cells at the back of the embryo, which are thrown into the rhythms of somitogenesis and allow the embryo to grow.

The set of interactions between the proteins—remember, genes only provide information for proteins—that creates those waves has come to be called the *somitogenesis clock* because of its repeated and precisely timed pattern of activity. A close look at these patterns of gene activity in individual cells reveals that they are oscillating and that the waves result from the synchronization of those oscillations in space. It will not surprise you to learn that the waves of gene expression seen in chicken embryos were later observed in other organisms and that they involved the same genes. The difference is the duration of the periods, the timing of the oscillations that follow the patterns of somite appearance. This is another example of the deep conservation of the activities of the genome and of the tools associated with specific processes.

Disrupt the waves, and you disrupt the appearance of somites and the patterns of ribs and muscles. In fact, many spinal cord defects in humans, including some instances of scoliosis, have been mapped to mutations in the genes associated with the somitogenesis clock. The waves are an integral part of the mechanism that generates the body.

Same genes, same waves, different periods, different numbers of somites in different animals: How does it work? If you take cells from the oscillating mesoderm, disaggregate them, and watch the somitogenesis clock genes in action, you will see that their expression oscillates and that their oscillations match the timing of the species they come from. The oscillations are intrinsic to the cells. Slowly they synchronize their activities, and after a while one can see persistent waves across groups of cells in the dish. Oscillations are an emergent property of a gene regulatory network, but the waves are an emergent property of the interactions between the cells in the tissue.

The level of the body axis at which the waves break is a fixed point from where they start and appears to be determined by other signaling systems that by now will be familiar to you: Wnt and FGF. So in somitogenesis we have a cell-intrinsic system of gene activity regulated by protein interactions that is coordinated at a higher level by signaling systems that shape space. The role of the cell in this central process is further emphasized by the observation that if you put the genes and proteins of the human in a mouse cell and ask it to engage in somitogenesis, the tempo of that network is that of the mouse. The cell has subverted the genes to adapt their action to the species they belong to. Once again, the cell rules.

During the elongation of the body, other kinds of mesoderm are also laid down and associated with the expression of *Hox* genes. Notching a regular pattern along the anteroposterior axis to make the buds for fore and hind limbs, the kidneys, the gonads, and the genitalia, they presumably operate under the control of similar timers. There is a simplicity to the iterative structure of somites, but not all of cells' self-organizing processes are quite as straightforward. To see the complexity they are capable of, we need only look at the making of a heart.

Your heart is a chambered box connected to the rest of the body by tubes that go in and out of its cavities. The two upper chambers, the *atria*, receive blood from either the circulatory system or the lungs, while the two lower chambers, the *ventricles*, pump blood out.

The vena cava, the pulmonary artery and pulmonary vein, and the aorta are the plumbing system, connecting the chambers with the body. After having fed the body with oxygen and other vital materials, exhausted blood is pulled along a network of capillaries and veins to the vena cava, two big pipes that feed into the right atrium. Once the right atrium is nearly full, its muscles contract, squeezing the blood down into the right ventricle. In turn, once the right ventricle is nearly full, its muscles contract, pushing the blood into the pulmonary artery to deliver it to the lungs, where it can pick up oxygen. After the blood has made its circuit of the lungs, it flows into the pulmonary vein to be delivered into the left atrium. When the left atrium is nearly full, its muscles contract, squeezing the blood into the left ventricle, and that chamber's muscles contract, pushing it into the aorta to distribute oxygen around the body. All of these muscle contractions are actually controlled by a part of the right atrium that receives electrical signals from the nervous system and shocks the two atria, causing them to contract in tandem. The electrical signal travels across the heart muscle to an area between the two ventricles, which causes them to contract, with great force, in tandem, to ensure that the blood can reach the entire body. This is repeated about one hundred thousand times each day, every day of your life. Out-of-sync signals and contractions create fast, slow, or arhythmic heartbeats. Mistakes can be fatal. The heart is a very sophisticated pump plumbed to the rest of the body, but it is what keeps you alive.

Of all the organs that develop in animal embryos, the heart is the first one to kick into gear, starting to beat very shortly after the dance of gastrulation is complete. At that stage it's still only a bulbous tube composed of cells of the mesoderm. In fish, the heart doesn't develop much beyond that, with the cells organizing into only one atrium and one ventricle that pump together to provide enough force to get the blood to flow throughout the body. In frogs, the heart cells go so far as to build three chambers, two atria and one ventricle, so oxygenated and deoxygenated blood can get mixed together. In mammals, which all share a four-chambered heart like ours, the cells start from the same tube, bending, folding, and fusing palisades of cells to create not only doubled upper and lower chambers but also intricate valves with two or three flaps that help to keep the blood flowing in one direction from chamber to chamber within the heart. The structure also needs to be connected to the veins and arteries that are emerging around it. In some ways it is a plumbing job. Most of this engineering and remodeling work happens while the structure is pumping blood through the embryo. It is amazing that so much effort goes into building what is essentially a pump.

More so than the brain, the mammalian heart is a pinnacle of natural engineering. Yet, oddly enough, between initially being a bulb of mesoderm cells and finally becoming a four-chambered marvel, it lingers for a period in a very different form. For part of their development mammalian embryos feature a two-chambered heart reminiscent of a fish's. A fish's complete heart and a mammalian embryo's heart in the making are not exactly the same, but the resemblance is telling even so. As von Baer would have said, the likeness speaks to some relationship, some throwback to a shared moment in our evolutionary history, which raises deeper questions about our place in nature.

The end of somitogenesis signals the end of the process of embryogenesis. By now, the heart has reached its final organization, and the different organs and tissues contain most, and in some cases all, of their functional components. Up to this point the story has been of cells rolling down the Waddington landscape, multiplying and using positional information to allocate themselves to specific fates. Next comes a more mysterious phase.

We know what happens to the embryo—the seeds of every organ grow proportionately to the size of the body—but we have little idea how this works. We know only that cells, as a group, "know" how much to grow and when to stop building, and this is why each organ has a defined size or, more surprisingly, your two arms have pretty much the same dimensions, which are different from mine, even though they have developed and grown independently of each other. We do not understand how this happens.

The one thing we know for sure is that cells continue to pull the strings.

THE POLITICS OF EMBRYOS

So far, gastrulation has been a story of cells, signals, and the structures that emerge during development and evolution. But during the beginning of the twentieth century, it became a political story as well. Ernst Haeckel is in large part responsible for this development. He was an excellent zoologist and embryologist who became one of the biggest champions of Darwin's theory of evolution. A gifted generator and communicator of scientific ideas, Haeckel illustrated his books with eye-catching vistas of the natural world, delivered scintillating talks at public lectures, and coined many words that remain in use today in the biological sciences, most notably *gastrulation*. His views about development and evolution were a persuasive ideological cocktail. At times, however, his eagerness to persuade led him to overreach.

The study of embryo development offers an almost tantalizing amount of visual proof of similarity across species, and many scientists have found it difficult to resist drawing conclusions based on visual similarity without experimental confirmation. Aware of von Baer's work, Darwin pointed in his *On the Origin of Species* to the significance of the similarities between early embryos, as well as their divergence later in development. He marshalled these facts among the evidence for his argument that animals, including humans, share a common ancestor.

Haeckel sought to express the same point, but he blended illustrations and words in a more dangerous, even self-defeating manner. In particular, in one of the illustrated plates in the edition of his book *The Evolution of Man*, published in 1874, he included a chart showing early stages of embryonic development at the top progressing to later ones at the bottom. Then he tendentiously ordered the menagerie of embryos that had been studied up to that time and populated the chart's columns, placing fish in the leftmost column and human beings in the rightmost one, suggesting that throughout evolution, organisms traveled along a road, from simplicity to complexity, akin to how they built up from simple balls of undifferentiated cells to complex organisms during development. This was already conjecture, but then he made another unwarranted leap.

Like von Baer, Haeckel pointed out that the younger the embryos, the more they resembled each other and suggested that, as animal embryos developed, they traversed the natural history of animal evolution as if climbing a ladder. Fish were a rung below frogs, frogs were a rung below newts, newts were a rung below chickens, and so on. Haeckel said that each stage represented evolutionary "progress," where organisms added improvements to the characteristics already built into "lower" species. If an embryo had gill-like slits early in its development, it really was a fish at that stage, not just fishlike. If later it exhibited a tail, as all human embryos do, this was because at that moment in our development we're actually a fish or monkey. He encapsulated this hypothesis with the mantra "ontogeny recapitulates phylogeny"—meaning that the embryonic development of an organism repeats the evolution of that organism.

From the beginning, Haeckel's drawings engendered suspicion. In the late nineteenth century, photographic evidence was not yet commonplace in science, and naturalists and physicians relied on drawings to illustrate their findings for readers. This granted them a degree of poetic license, which Haeckel stretched to the limit, sometimes drawing embryos to suit his ideas in a manner that did not match reality. For some drawings he copied from others' sketches rather than observing the embryos himself. In some he labeled a drawing of a chicken or dog embryo as human. In some cases he repeated the same drawing while stating that it was a series of different images of different animals at the same stage of development.

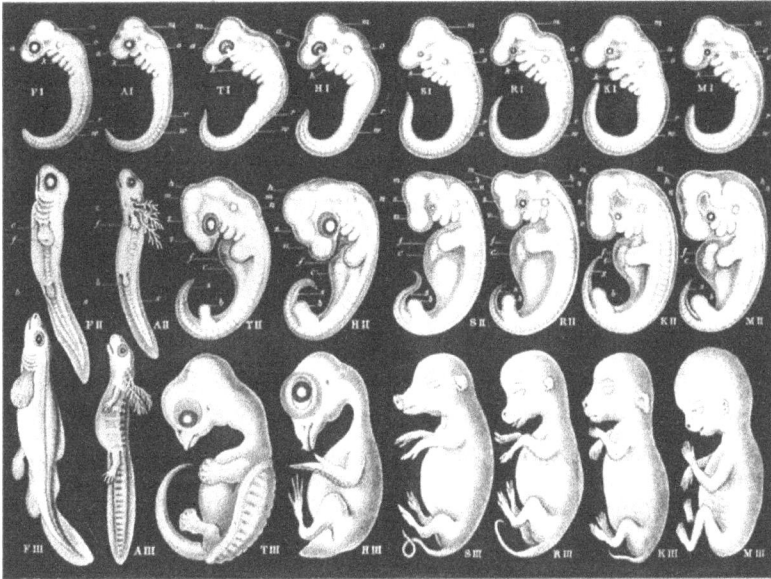

FIGURE 24. Ernst Haeckel's comparative drawings of vertebrate embryos during development. From *Anthropogenie oder Entwickelungsgeschichte des Menschen,* 1874.

Haeckel's "fast and loose" approach to the illustrations exposed him to the censure of his fellow academics. He publicly admitted what he'd done and corrected the images in future printings, but the damage was done. Haeckel would become associated with fraud. Worse, proponents of so-called intelligent design, which posits that God's divine hand created every organism just as it is, continue to use Haeckel's drawings as "proof" that all evidence for evolution is concocted. Search today for "Haeckel embryos evolution" online, and your top ten results will almost certainly return words like *fudged* and *lies* as well as several antievolution and antiscience websites.

It is unfortunate that, by stretching reality, Haeckel's drawings have attracted so much criticism and ideological attention. The controversy has cast a shadow over central facts in both development and evolution that deserve to be studied in detail—most especially, why vertebrate embryos *do* all exhibit similar, though not identical, organization at the end of gastrulation, the moment at which the *Hox* genes are unzipped and become available for the cells to use in building the organism's body. Analogous structures can be seen in many invertebrates too. Something must drive these universal patterns of development across animals, including humans. Decades before DNA was discovered and unraveled, embryos hinted at our common ancestry with other animals.

Where does this leave us? How can we make sense of what and who we are? When Haeckel placed primate embryos in the far-right column of his developmental chart, he asserted a nonexistent scale of progress and perfection, with us at the top. Including humans in a biological classification system almost always causes headaches because, for the most part, we accord ourselves a moral and intellectual status above other animals, which, in some cases, we ought to question. Are we, as Genesis puts it, "created in the image of God" or, as Shakespeare wrote, "the paragon of animals, the quintessence of dust"? Or might we be something else? The only way to know would be to observe our own origins in the womb itself.

six

HIDDEN FROM VIEW

W HAT DOES A BABY LOOK LIKE BEFORE FULL TERM? THIS question becomes inescapable for those who undergo the scary but not uncommon experience of a premature birth. My family joined this group when my second child, Daniel, was born prematurely, at twenty-eight weeks of gestation. The ordeal had a noticeable effect on his grandmother as she met him for the first time.

When I picked my mother up at Heathrow Airport in London, she was unusually quiet. The drive to our local hospital took two hours, and yet she hardly uttered a word—odd, as there was a lot of news to catch up on, and we usually spent such journeys in conversation. At the hospital, she remained silent as we walked to the premature baby unit. There, she circled the incubator a few times, looking intently at her grandson, before turning to me and saying, "He has everything!"

I finally understood the depth of her disquiet: she had not known what to expect of a baby born so many weeks before full term. She was worried that he would not be "normal." For her, like so many of that generation, what happened in the womb during pregnancy was a mystery shrouded in wild imagination and fear.

My mother was relieved to see that Daniel was fully formed but not fully developed. Although it is well in the third trimester of pregnancy, twenty-eight weeks was then considered to be the borderline between a life with or without complications; in too many cases, it was also the borderline between life and death for a newborn. His brain was still growing at a very fast pace. His lungs weren't developed enough to allow him to breathe on his own, and he had not yet developed an instinct to suckle; he still depended on getting nourishment through the exchange of blood with his mother, my wife, Susan. He spent the next two months in the premature baby unit, where the staff nursed his continuing development until he gained the strength to go home. Under the supervision of the nurses of the unit, Susan, his sister Beatriz, and I would all spend many more months navigating the risks and challenges of his premature birth, and we shall always be grateful to the UK National Health Service for what they did for Daniel and, through achieving his full development, for us.

Today, he's a healthy adult. We know that some aspects of his life have been shaped by finishing his development in a hospital incubator and crib rather than Susan's womb. But much of who he is was created weeks before his premature delivery into the world. This was the handiwork of cells, and much of it was laid down before the eighth week of gestation, when a human embryo has grown to a length of around half an inch, and all of the elements of a recognizable baby are in place—the arms, the legs, and a head with outlines of eyes, nose, and mouth. From this moment we describe it as a fetus. Prior to this transition, the embryo's cells are busy lining themselves up into small cliques, the seeds of organs and tissues, jostling for space and creating the links needed for function. As in other animals, gastrulation is responsible for laying out this organization, but even earlier another essential step takes place. The first hundred or so cells arising from the zygote commit to one of two paths: generating the embryo or supporting it. As mammals, we need to create an attachment to our mother—one that will provide food and protection. As a matter of fact, there are several connections, and they all arise early in development.

MAMMALS: THE HIDDEN EGG

Every kind of animal is unique in the way it is made. However, since the days of Carolus Linnaeus, we've grouped organisms into classes, and within those classes are animals whose cells share strategies for weaving together the body. As we saw in the previous chapter, fish and frog embryos develop rapidly from translucent eggs fertilized outside the mother's body. We know a great deal about how this process unfolds since we can watch the emergence of a fish or frog from the initial fusion of sperm and egg up to the cellular dance of gastrulation, when the body plan forms, without much trouble. We can also do it with birds: carve a small window in an eggshell, and you can become a modern Aristotle observing the emergence of a chicken from a zygote on a daily basis.

Mammals are different. We are a small group of animals; out of 8.5 million species on earth there are only 6,000 of us. We all share a somewhat strange mode of development: as embryos we grow inside our mothers. For this reason, mammalian fertilization and development are naturally hidden from view, happening deep inside the mother's womb. As a souvenir of this form of development, we carry with us a mark that dates back to the time we spent attached to our mother: our navel, or what Aristotle called "the root" of the belly.

The navel is a remnant of the umbilical cord, which is generated very early in development. It connects the embryo to the placenta, an accessory organ that anchors the embryo to the womb and creates a cellular system for the transmission of nutrients and disposal of waste. When, in the course of evolution, these structures and connections emerged in mammals, they appeared alongside a reduction in egg size. As we saw in the scientists' difficulties transferring the processes of cloning from frogs to sheep, mammals' eggs are much smaller than those of other animals—so small that, while they contain some yolk-like nutrients to fuel the first weeks of development, they don't have a large enough store to sustain mammalian development all the way to completion. While some egg-laying mammals, like the platypus, do have much larger eggs, they are a rare exception. To build mammals, cells had to invent a way to feed off the mother.

Only relatively recently in the history of science have humans grasped how this process works. In fact, because most mammalian eggs are so small, for most of recorded history it was assumed that perhaps mammals didn't come from eggs at all. From at least Aristotle's day until the seventeenth century, it was supposed that babies were generated when semen was planted in the fecund "soil" of a woman's womb, though how body tissues arose was a matter of heated debate. Because fish, frogs, and birds all clearly came from eggs, as we saw in Chapter 5, the king's physician, William Harvey, pushed the very reasonable idea that mammals must come from eggs too. However, mammalian eggs proved elusive. Then, in 1827, around the same time that he forgot to label his early embryo specimens and discovered he couldn't distinguish them from each other, Karl Ernst von Baer dissected the ovary of a dog that had gone into heat:

> I discovered a small yellow spot in a little sac, then I saw these same spots in several others, and indeed in most of them—always in just one little spot. How strange, I thought, what could it be? I opened one of these little sacs, lifting it carefully with a knife onto a watchglass filled with water, and put it under the microscope. I shrank back as if struck by lightning, for I clearly saw a minuscule and well-developed yellow sphere of yolk. . . . I had never thought that the content of the mammalian ovum could look so similar to the yolk of a bird's egg.

Here was proof at last: mammals had eggs. They were simply incredibly small.

Although it's the largest cell in the body, a human egg is a mere one- to two-tenths of a millimeter in diameter, about the same as that of a strand of hair and five hundred times smaller than the size of a chicken egg. In theory, that's large enough to be seen with a naked eye—but you must know where to look, as von Baer realized belatedly. After his discovery he later lamented that on the rare occasions in the past when he'd been able to examine a pregnant corpse, he hadn't known that mammals' eggs were "so very small," and so he'd missed them.

Human eggs are not just tiny; they are fragile, fleeting things. After each egg is released from an ovary, it lives for just twelve to twenty-four hours, traveling along a fallopian tube toward the womb, where it will disintegrate unless sperm has fertilized it. Even after egg and sperm fuse into a zygote, the resulting cell is not much easier to find: that single cell divides and multiplies while rolling freely around the womb until around Day 7, when the resulting mass of cells attaches itself to the uterine wall.

From this moment, it would seem that the making of a new human is definitively underway. The details of what cells do next, particularly during gastrulation, have extraordinary bearing on the person who will be born, but this is the moment when, for ethical reasons, scientists cease to be able to do very much at all to study the forming fetus. Instead, they must resort to studying our fellow mammals to uncover how the massing cells sketch out a body plan after they settle themselves in their new home to gestate.

TOOLS OF MICE AND MEN

There are inescapable limits on what scientists can learn about the exact details of gastrulation in humans due to the sensible prohibition on experimenting with human embryos. So most of what we've learned about human development has, until very recently, come from observing the behavior of cells from other mammals. Rabbits, pigs, and sheep have proved to be very useful in this research, but the reigning favorite in laboratories is unquestionably the mouse.

Why mice? Because, in addition to being small, breeding well, and developing quickly—which, as we have seen, has been crucial in mutant hunts—they share 97 percent of their DNA with humans. Not only that, but among the 2 percent or less of human genes that code for proteins, the so-called toolbox of the genome, mice share a remarkable 85 percent with us, with long stretches of the mouse genome featuring the same or a similar order and number of genes, lining up remarkably well against our own. For these reasons, mice have been key to understanding many aspects of human biology and teasing out specific genetic mutations that affect development, as we saw in Chapter 1. They've also been especially important in unraveling our understanding of the first weeks of mammalian development.

Scientists can grow mouse zygotes successfully outside the womb in culture dishes for two or three days, until the embryo is implanted in the uterus. From this we have learned that, after fertilization, the mammalian zygote divides very slowly compared to those of other animals. While a fruit fly zygote generates about 6,000 cells in just two hours, at which point gastrulation begins and takes about one hour, the mouse zygote divides only about two or three times a day, generating around 240 cells after four days, when it will implant.

Still, though the timing and the number of cells produced vary, much else that occurs in these early divisions is the same in all mammals. After the zygote has divided three or four times, it has assumed the form of a tight aggregate of cells, called a *morula* (*mora* means *berry* in Latin, and the cell possesses a berrylike shape). Some of the morula's cells are positioned inside the mass; others are on the outside. Up until a certain point, every one of those cells could give rise to a whole organism. Scientists have found that any cell that is taken out at this early stage maintains the ability to divide and multiply just like the original zygote. Split the group of cells into two or three or four, and the smaller cell clusters can resume dividing and multiplying as the original morula had been doing when it comprised the same number of cells. Identical twins most likely arise from a splitting of the zygote at the two-cell stage or of the morula into two aggregates. By the same measure, if you were to add an additional cell or mass of cells, the morula would accept them and proceed as if it were further advanced in its process of splitting and growing; chimeras likely arise from a fusing of multiple aggregates.

Decisions about the mammal's cellular destiny lie ahead. But before the morula develops further and affixes itself to the uterine wall, the first step in assuming defined roles for making the embryo is taken when cells focus on building a nest for the embryo that will be sewn within the womb. The cells on the very outside of the morula start a script of activity that changes their appearance and function, creating a "wall" of cells. This palisade, one cell width in thickness, is the precursor to the placenta, which is why these cells are collectively called the *trophectoderm*, from *tropho*, meaning *nourishment*, reflecting the role that this structure will have in feeding the embryo off the mother. The cells on the inside of the morula remain just as they are, undecided, for the time being.

Embryonic cells develop by using the toolbox in the genome. The cells of the trophectoderm do this by using a protein called CDX2, which controls the structure of the cellular palisade and places a collection of pumps within it. These pumps control whether liquid is allowed into the morula. When they're turned on, liquid flows in and creates cracks in the spaces between the cells in the morula's interior—a biological process akin to fracking, where water and chemicals are blasted into rock causing it to fracture and allow liquid to pass through.[1] Once these pumps have done their work, a structure takes form: the palisade of the trophectoderm now surrounds a liquid-filled cavity with a mass of about 100 to 150 cells gathered against one area of the wall.

Amazingly, if you take a cell that's on the inside of the morula and move it to the outside before the biological fracking process is completed, the relocated cell will recognize its new position, start producing CDX2, and join in the pumping process to help create the liquid-filled cavity. Move a cell that's on the outside of the morula to the inside, and it will stop producing CDX2. This demonstrates that cells are actively assessing whether they lie on the outside or the inside of the morula and acting to control gene activity accordingly.

Once the liquid-filled cavity is formed, the mass of cells inside the trophectoderm wall begins to change. It is time to settle on an occupation at last, though not all will end up generating the embryo even at this stage. Not just position but precise proportion will now become vital. One more decision takes place at this stage in the inner cells: one population becomes a group of cells that will feed the embryo until the mother takes over and, for this reason, is called the *primitive endoderm* or *primitive gut*, because the structure that it will produce—the yolk sac—will feed the embryo until the placenta takes over. The rest will become the embryo.

The ensemble is called the *blastocyst*, and now all is set for this society of mammalian cells to implant itself in the womb and begin the dance of gastrulation.

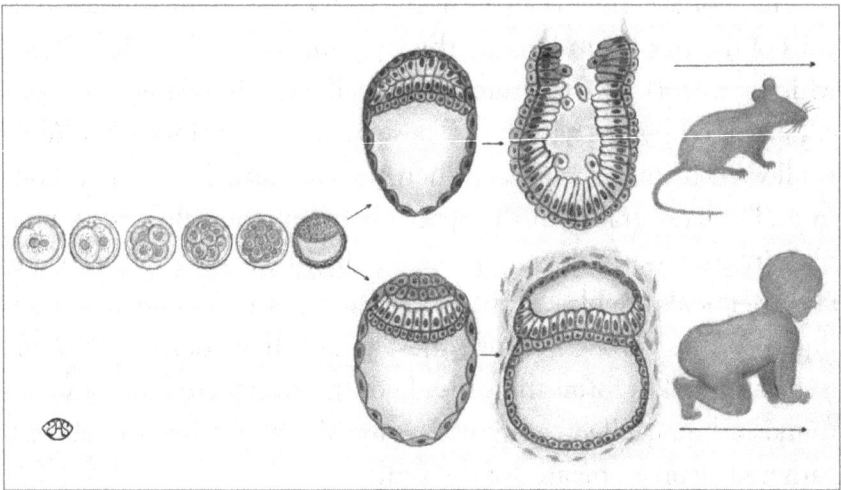

FIGURE 25. The early development of mice and humans is initially very similar. After fertilization, the cells divide, multiply, and segregate the extraembryonic tissues, which can be clearly seen in the blastocyst: the trophectoderm outside and the primitive endoderm, lining the cavity side of the embryonic cells, tucked between these two cell types. The embryo is ready to implant into the uterus, and now differences emerge between mouse and human as the embryos prepare for gastrulation: the mouse cells organize into a cup shape, while human cells form a flat structure.

This way of talking about what is happening in cells differs greatly from the language used by geneticists. In their view, genes are the bosses, the engineers, the drivers of the events that decide when and where something happens. Yet, as we can already see, the cells are the ones who count and read signals from their neighbors and assess their position in the community, sensing not only the chemical signals they exchange with each other but also the physics of geometry, tension, pressure, and stress within and across the group. It is the cells that, in agreement with their position and surroundings, reach into the genome to turn on or turn off the tools they need to build tissues. We know that the cells are in charge because they are capable of changing what they're doing if they're moved from one location in the blastocyst to another or if their society of cells is split in two.

Throughout these early stages, the cells in this ensemble that is creating the blastocyst signal to each other using FGF, Wnt, and Nodal, signals we have seen before, and a new element of their repertoire, a signaling system called Hippo, which conveys forces and information about position and numbers of cells. The genome supplies the tools, but the cells do the work.

ATTACHED TO OUR MOTHERS

Though embryonic development proceeds along similar lines in all mammals through these early stages, the moment the blastocyst implants itself, differences arise. The blastocysts of most mammals use the trophectoderm to attach themselves to the mucous membrane that lines the uterus, the *endometrium*, sitting on its surface. Some, like those in cows and horses, attach after gastrulation! However, in a small set of mammals, including primates, blastocysts get a stronger foothold early, settling into a fold within the uterine wall before gastrulation. There are other differences as well: mice and other rodent blastocysts line up in a cup shape, while the blastocysts of all other mammals, including humans, form a flat disc. These shapes are the handiwork of the trophectoderm, which either pushes down on the embryonic cells (to make a cup) or stretches them out (to

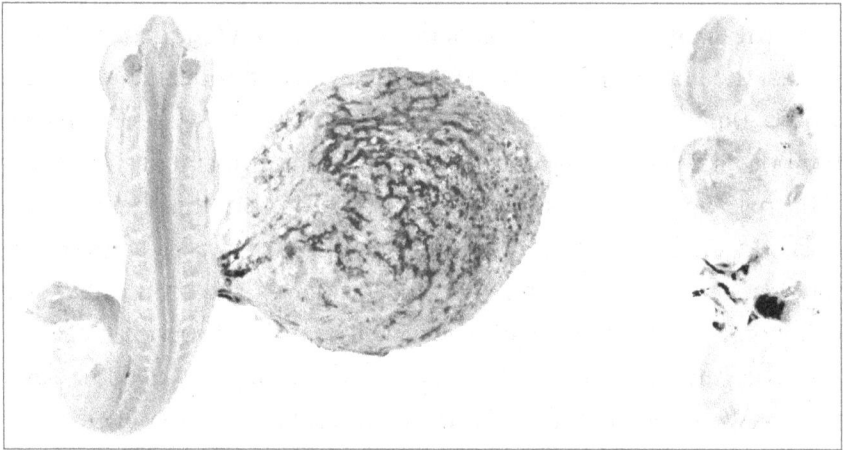

FIGURE 26. Back (*left*) and front (*right*) views of a human embryo at twenty-eight days with yolk sac attached. The view from the back displays the somites on either side of the spinal cord. In the front view, the primordia of the eyes can be distinguished in the head, and a primitive heart is tucked underneath.

make a disc). The cells of the trophectoderm do this not by increasing or decreasing the number of cells but by changing the tension within and between cells.

Many aspects of human development are more familiar today than they were just a few generations ago, when my mother and her peers grew up. In large part this is thanks to technologies like ultrasound and in vitro fertilization (IVF). Credit is also due to the Swedish photographer Lennart Nilsson, who published a stunning photo-essay in *Life* magazine in 1965. Titled "The Drama of Life Before Birth," it gave the general public its first opportunity to witness the transformation of an aggregate of cells into a human being, from the first appearance of eyes and limbs in an embryo to a fetus learning to suckle by sucking a thumb while swimming in amniotic fluid. Suddenly, human embryos were visible to everybody—and from a very early stage they looked surprisingly human.

For much of human history, people hadn't considered pregnancies in terms of embryos. Indeed, up until von Baer confirmed their existence in 1827, mammalian eggs were merely a hypothesis. Most religions and laws drew a line demarcating the moment when developing life became human at what was then called the *quickening*—the moment when the mother first feels the stirring of a child in her womb. This typically occurs between the eighteenth and twentieth weeks of gestation, well after the embryo has developed into a fetus and around the time that the pregnancy might be starting to show. This is why the first laws against abortion, passed in the early nineteenth century, made it illegal to perform an abortion only after quickening.

Many of the embryos and fetuses photographed by Nilsson came from miscarriages and aborted pregnancies and were staged as if they were still viable. The developing embryos and fetuses were presented alone, floating in space like an astronaut outside his or her capsule; in some of the images, they were enveloped in a fluffy gauze meant to evoke the comfort and safety of the mother's womb. Though the images included the umbilical cord, few hinted at its purpose: maintaining the essential, life-sustaining link to the placenta and mother. This approach, portraying the fetus almost as an independent creature, has become standard in depictions of human development to this day, such as the ones found in apps for expecting parents. Unfortunately, pro-life groups have seized on such images to support the idea that a unique and fully fledged human being, with the capacity to survive on its own, exists from the earliest days of development. That's just not true.[2]

Fundamentally, so long as we are in the womb, we're dependent on our mother, and efforts to portray the fetus as a separate person simply make no sense. The relationship between the two is intimate and the umbilical cord provides a good example of this. Beyond its obvious role of supplying nutrition and oxygen to the embryo or fetus, the umbilical cord is also a two-way conduit that can allow exchanges in either direction. Any cells that travel through the blood can travel from the mother to the embryo, and vice versa. As a result, an embryo gets some cells from the mother, and the mother receives some cells from the embryo.

Early in the pregnancy, the embryonic cells that travel over to the mother might not yet have been fixed on a path to becoming a particular organ or a tissue; they're *multipotent*, meaning they can give rise to many but not all cell types. Because their fate, form, and function are still malleable, they can be enlisted to help build or repair damaged tissues anywhere in the body—including in the mother's body. Studies in mice have shown that cells from the embryo can indeed repair damaged maternal tissue, and there is evidence that the same might happen in humans.[3] This is detected by the presence of cells from male embryos in tissues from the mother. The phenomenon is called *fetal cell microchimerism*, as a mother becomes a chimera not of genes but of cells from the embryo. One could view this phenomenon as an instance of the embryo helping the mother in the cooperative nature of cells.

IMITATING NATURE

Ever since von Baer confirmed that mammals have eggs, it's been clear that making a human requires combining an egg with sperm and letting the newly minted cell run, so long as it can access a womb to feed from. Indeed, this is how to make just about any mammal, and it's essentially what Robert Edwards, Jean Purdy, and Patrick Steptoe did when they first successfully developed the technology of in vitro fertilization for humans, creating a zygote in the lab and implanting the resulting aggregate of cells in a mother. The result was a healthy baby girl called Louise Joy Brown, born in the summer of 1978 in Oldham, UK.

This technique has been in use for decades, and yet the notion that a human egg can be fertilized in a dish is still highly experimental. Only around 35 percent of attempts work. And it's through the history of doctors and scientists trying—and failing—to start making a mammal outside the womb that we can get a better sense of how much the surrounding environment matters to the cells that created us.

Back in 1944, Miriam Menkin could sense she was getting close to something big. Six years earlier, having previously worked with Harvard biologist Gregory Pincus on research to create "fatherless" rabbits, she'd been hired to help John Rock at Boston's Free Hospital for Women, who at the time was seeking a way to help infertile couples have a baby. No one knew if this was possible, but her mentor, Pincus, seemed to have fertilized rabbit eggs in a dish; he even claimed that a bunny born in his lab in 1935 was the result of an egg being fertilized in vitro. Why shouldn't the same technique work for humans?

So, Rock would ask women having a hysterectomy at the hospital under specific medical conditions if they'd be willing to have their surgery just before they were supposed to ovulate. This would ensure that a portion of the eggs in the removed ovaries were mature and primed for fertilization. After the surgery, Menkin would extract the eggs from the ovaries, combine them in a dish with some sperm for thirty minutes, and see what happened. Week after week, she saw nothing. She changed the conditions in the dish and considered differences in the sperm, but still she was unsuccessful.

Then one day, having had to care for her own child, she forgot the time and left the eggs and sperm together longer than usual. She returned to the lab and found that a zygote had formed. The resulting cell had even progressed, dividing to generate two and then three cells, before they stopped. It was the first time that a human egg seemed to have been fertilized outside the womb—and it suggested that human cells worked to their own schedule. There are doubts that she had observed real IVF—even Pincus's 1935 IVF bunny seemed increasingly likely to be the result of an old-fashioned pregnancy. But the results created momentum on the way to the objective.

Then, in 1959, one of Pincus's colleagues at the Worcester Foundation of Experimental Biology, Min Chueh Chang, made a remarkable breakthrough: he fertilized the egg of a black rabbit with the sperm of another black rabbit in a dish, then implanted the resulting aggregate of cells into a white rabbit. When the white rabbit gave birth to a black bunny, it was definitive proof of a live IVF birth; because black coats are a recessive trait in rabbits, there was no way that the white surrogate could be the mother. With this clear proof that a zygote created in a dish could develop into a fully functional and developed organism once placed in the natural environment of a womb, the foundations were in place to work out the practicalities for human IVF pregnancy.

First, scientists would need to understand a great deal more about how eggs functioned. Enter Robert Edwards, who had grown up in a working-class family in Yorkshire and served in the British army before studying biology at Bangor University and then the University College of North Wales.[4] He earned a PhD in animal genetics and embryology at the University of Edinburgh under the mentorship of none other than Conrad Waddington, focusing his attention on the effect of chromosomal abnormalities on mouse development. To understand these, he began to zero in on the workings of the egg—how it matured, how it primed itself for fertilization, how it interacted with sperm to create a zygote, and what role it played in the development of the organism. In the process, he overcame limitations on the number of eggs mammals produced by inventing a way to trigger superovulation, which would come in handy later on.

In 1963, he transferred his study of the initial stages of development in eggs and sperm from mice to humans. Having joined the department of physiology at the University of Cambridge, he hired Jean Purdy, a former nurse who had conducted research on tissue rejection, as part of his lab team. Like Rock and Menkin before them, Edwards sought human egg donors, which proved almost impossible to find. Edwards visited hospital ward after hospital ward, mainly in London, begging for samples of uterine lining from which he hoped to scrape eggs. He got some, but not nearly enough.

Progress was excruciatingly slow until Edwards met Patrick Steptoe, an obstetrician at Oldham General Hospital, near Manchester, who had pioneered keyhole surgery techniques for procedures such as sterilization. Edwards hoped that Steptoe would be interested in using his surgery skills on women willing to donate eggs but unwilling to have a hysterectomy. Steptoe went above and beyond Edwards's expectations. Passionate to end the pain of infertile couples, the obstetrician saw in Edwards and Purdy's work the potential to make IVF a reality and joined their team.

The trio began their collaboration to create a baby in a dish in 1968 and in 1969 reported the successful fertilization of a human egg in vitro for the first time since Menkin's accidental zygote in 1944. Over the next year, they got the zygote to divide and multiply into sixteen cells, and a year after that, they created two blastocysts, complete with the palisade of the trophectoderm, the liquid-filled cavity, and the mass of cells inside differentiating based on their position. All that was left was to implant a blastocyst at this stage into a womb, where the cells could continue to develop on their own.

To move on to the final stages of IVF fertilization, they would need funding, and where matters of funding are concerned, politics often becomes an issue. The team had permission to set up a clinic near Cambridge to bridge the gap between their laboratory research and medical treatment, but the Medical Research Council (MRC) denied them a grant to fund it.[5] The panel, and others, voiced concerns that implanting a lab-created blastocyst in a womb might induce abnormalities in whatever embryo might result. How would the public react to tax dollars funding such a thing?

The team scrambled to obtain private funding and shift their operations to Steptoe's clinic. The eggs were collected in Oldham, then packed up and taken to Cambridge, where they'd be put in a dish with some sperm in different conditions to create zygotes. Once they had succeeded in dividing, creating sufficient clusters of cells, they'd be carted back to Oldham to be implanted in the mother—a twelve-hour round-trip drive with the most precious of cargo.

Perhaps the MRC panel's concerns were not off the mark. At the time, there were no sophisticated ultrasound scans available to check if development was proceeding normally. If the experiments were being attempted today, not only would they have been denied funding—they likely would never have been allowed at all. But after 495 treatments with 282 couples, on July 25, 1978, Louise Joy Brown was born. Today, IVF is a multimillion-dollar industry, bringing joy to many and hope to many more.

Scientists' ability to fertilize eggs in vitro and grow a zygote into a cluster of cells ready for implantation has given us significant insight into how the embryo comes into being, particularly how development differs in humans versus other mammals. Critically, while mice are an excellent reference for us to understand the general principles of development, some details are specific to humans—and in early development, these details matter greatly. Mice and humans differ not only in the time it takes to make a blastocyst—four days for a mouse embryo versus seven days for a human—but also in the way the cells communicate with each other. And despite having essentially the same genomic tool set, mouse cells and human cells wield these tools in subtly different but important ways. For example, in mice the FGF signal is essential to specify the primitive endoderm, but this is not the case in humans.[6]

The details matter—especially in the early stages. Considering the fact that most pregnancies are lost during the first week after fertilization, the cells' maneuverings during this period are clearly fundamental to the organism's very existence.

DRAWING THE LINE

We humans tend to think of the world in individual terms, particularly when our offspring are concerned. It's more and more common for parents-to-be to pick out a name for their fetus, muse about its gender, and imagine its future characteristics long before it has even a reasonable probability of survival. But science has a way of defying our cultural expectations.

The majority of natural in vivo pregnancies prove nonviable, and, as we have seen, IVF pregnancies have a 35 percent rate of success. For this reason, a woman undergoing IVF is given a cocktail of hormones to trigger superovulation, meaning that instead of one egg being primed for fertilization and released during ovulation, two or three eggs are. After these eggs are harvested, they are fertilized while they are mature, and each of the resulting zygotes is allowed to begin dividing and multiplying. When they reach the blastocyst stage and are considered ready to undergo further development, one or two are implanted, while the rest are frozen in case they are needed for future pregnancy attempts.

But freezing these groups of cells raises a host of questions. What is a frozen blastocyst's legal status? Can it be owned, and if so, does it belong to the donors of the egg and sperm or to the scientific team that created the zygote? If an IVF pregnancy is successful, could any surplus sets of cells be ethically used for purposes other than IVF? When does the clump of cells become an individual—at the blastocyst stage or only later? And if a blastocyst is an individual, does it have rights? The last question was—and is—particularly important, because while there is no doubt that these cells are human, there remains no consensus on when an aggregate of cells constitutes an individual human being. The invention of IVF therefore forces us to confront an age-old question: If there is a soul, when does it come into our bodies?

In 1982, a committee was set up in the United Kingdom under the stewardship of the philosopher Mary Warnock to consider these questions—though not the one about the soul—and make recommendations about how to handle research with human embryonic cells going forward. The twenty-one committee members represented a range of disciplines—including philosophy, law, religion, and medicine—and included just one scientist, developmental biologist Anne McLaren. Over the course of two years, the members considered evidence and debated the issues created by the development of the technology. They decided to set aside the bigger question, *When does life begin?*, agreeing that this was "a matter of belief as much as science." Instead, they chose to take the perspective of the embryo and the individual and consider what "degree of protection, if any" each should be given. In fact this proved not to be a simple issue either, since even the question of when what we have called a morula, a blastocyst (the name for a mammalian blastoderm), embryonic cells, an embryo, or a gas-trulation embryo becomes an individual is highly complex.

The committee members gathered opinions from over three hundred organizations from a variety of religious traditions and social and human rights viewpoints. They solicited letters from the public, with nearly seven hundred people writing in. Catholics advocated for conception as the moment when the individual was created. Jewish and Muslim experts said that an embryo becomes a human around Day 40, close to the transition from embryo to fetus.

Some scientists suggested the turning point comes when the first sparks of the nervous system appear, since an organism would presumably start then to sense, feel, and think. Others pointed to the moment that the heart starts beating. Yet defining when a heart comes into existence is not a simple matter either. Just as a group of cells that will become heart tissue start to register a fluttering before that happens, cells sometimes begin signaling to each other with electricity before it is clear what they are destined to become. Clear distinctions simply do not seem to apply at this stage of human development.

The group's sole biologist, Anne McLaren, played a leading role in the final decision. Renowned for her experiments transferring developing mouse embryos out of and back into the womb and her knowledge of mammalian embryology, she was able to clearly articulate the intricacies of early human development to her fellow committee members, describing how groups of cells moved around and specialized as they became embryos, focusing on the process of gastrulation in particular. She showed how the organization of the rare early human embryos that had been saved and stored in medical collections related to what biologists had learned from observing the early embryos of other species. And she must have made her views as a biologist clear—as she did in a paper she wrote on where the individual came into existence. The paper is titled "Where to Draw the Line?" and some of its content were used in the committee's discussions.[7]

"If I had to point to a stage [of development] and say, *This is when I began being me*," she wrote, "I would think it would have to be [Day 14]," the moment at which gastrulation starts.[8] And around that day in human embryonic development, a literal line is drawn: it's when the primitive streak, the furrow ploughed in the field of cells that, you may recall, establishes the front–back axis of the body, starts to appear. The choice, though arbitrary, was no whim.

Why had McLaren pointed to this moment? Let's return to the magic of the dance that occurs in chicken embryos undergoing gastrulation. As we saw earlier, chicken blastocysts, like human blastocysts, form a disc shape after the sequence of cell divisions that follows fertilization. In the chicken, that disc contains thousands of equivalent cells. Then the dance starts at one point with the ploughing of the primitive streak. If you split the disc in half before this moment, cells in each half will form their own primitive streak, and two embryos will emerge. If you split the disc in quarters, you'll get four primitive streaks and four embryos. The primitive streak is a marker for a full body plan.

McLaren knew that the human blastocyst is also shaped like a disc, though it has far fewer cells and a different organization. Nonetheless, in the few cases where scientists had got a glimpse of these early stages of human development, they'd seen a furrow, a streak, resembling the primitive streak in birds. This happens sometime during Day 14 or Day 15 of development, the start of gastrulation in human embryos. Of course, things can go wrong during gastrulation too. It is believed that most instances of conjoined twins are due to the emergence and fusion of two primitive streaks or the splitting of one streak into two. Experiments with chicken embryos support this interpretation. And so, following McLaren's guidance, the Warnock Committee suggested that an individual human being came into existence around Day 14, or the appearance of the primitive streak, as this is the last moment at which splitting the group of cells into two will give rise to two individuals.

In 1984 they published the Warnock Report establishing the "14-day rule," the outer time limit for experimenting with human embryos, and this rule was enshrined into UK law in 1990. Several countries followed suit.

The ability to grow fertilized eggs in vitro until implantation, developed for IVF, opened new avenues for understanding how humans develop from a zygote, and over the last ten years technical advances have allowed labs to grow human embryos in culture right up to the limit of Day 14. Previous research confirmed that human cells maneuver themselves as they multiply and divide in very similar ways to other mammals. It allowed us to observe how long they take at each cell division, learning the distinct rhythms of humans' early cellular timer. We've learned to see how the shape and appearance of a blastocyst hint at its chances of surviving to term. In addition, the latest experiments have suggested that a small portion of embryos undergo changes in the dish that were previously only associated with implantation in the womb.

It is difficult to assess what this means. The embryos involved are usually not healthy and wouldn't have survived much longer, even if it weren't legally necessary to destroy them. Furthermore, given that over half of zygotes don't make it to gastrulation, it's not easy to know whether the defects we observe are due to the experimental conditions or part of the natural course of events. But it's only a matter of time until scientists find the conditions that will yield higher rates of survival. Whether the 14-day rule will then be allowed to stand without any qualifications is far from settled.

In any case, we know that the dance of gastrulation holds many more secrets of human development, even if we are not able to watch it happen in humans. In the pursuit of defining a zygote's rights as an individual, the Warnock Committee pushed gastrulation, the most important moment in making us the humans we will become, back into the darkest shadows. Recovering new clues from this informational black hole demands creativity, but perusing the annals of the history of science gives us hints at what we are missing and what we should use as a reference point.

HUMAN EMBRYOS

The moment where we lose sight of the process of embryonic development always occurs one week after sperm penetrates egg. At this moment the aggregate of cells that will give rise to a human being is wrapped in the trophectoderm, the layer of cells that later becomes the placenta. This is when IVF pregnancies are introduced to the womb, following the natural process wherein human blastocysts initiate their invasion of the uterine wall on Day 6 or 7 after conception. Once the blastocyst has burrowed in and gastrulation has begun, it's nearly impossible to extract the embryo to observe how it's developing—unless something goes wrong.

This is why, for many centuries, what we knew about human development came from collecting specimens of pregnancies ended through miscarriage or abortion. Some can still be seen in collections like the Royal College of Surgeons' Hunterian Museum in London or Le Grand Cabinet de Curiosités at the Jardin des Plantes in Paris. What you will find in those jars, if you choose to visit them, is striking.

For the most part, the fetuses and embryos display *abnormal* development, which at the time was deemed normal. Split skulls expose misshapen brains, split-open backs reveal broken spinal cords, fused legs create the illusion of a mermaid, and cyclopic creatures and conjoined twins feature frequently among the displays. It's little wonder that, in many cultures, the specimens that emerged from early terminations were deemed to be not human but rather lesser animals, ephemeral spirits, or lumps of flesh—nothing of human value. There was, nevertheless, shock value in them, and they were considered too unseemly for women to see. A poster for a display at a show in London in the late nineteenth century read, "Embryology, or the origin of mankind, from the smallest particle of vitality to the perfectly-formed foetus. For gentlemen only."

Among the collectors and traveling showmen were those who shared with contemporary scientists an interest in reconstructing early human development. One of the scientists was Swiss anatomist Wilhelm His, who set out to define "normal" human development in relief against this menagerie of malformed embryos and fetuses. To do this he first needed to gain access to a sufficient number of specimens to populate as complete a timeline of development as possible. With a focus on the first two months of development, he contacted doctors, midwives, and scientists to ask for uterine tissues from any failed pregnancies they encountered. He was also shrewd: knowing that specimens of human development were rare, and people were not always happy to give them up, he claimed to have the ability to discern what a normal embryo looked like and promised to interpret any abnormalities. As a further reward for providing tissue, he offered his sources—which encompassed professionals, mainly gynecologists and midwives, but not the mothers themselves—eternal credit for supplying a specimen by naming it after them. In this manner he helped establish a tradition that would continue well into the twentieth century, whereby the woman who bore the embryo or fetus was edited out of medical history.

Wilhelm His's collection eventually included twenty-five normal embryos at between 3 and 8.5 weeks' gestation, allowing him to publish the first methodical account of human development in 1885.[9] The specimens reinforced humans' links to other animals while refuting Ernst Haeckel's notion that we become human after passing through other animal forms while in the womb. His showed that from the zygote onward, we are and remain uniquely human.

His's work was carried on by one of his students, Franklin P. Mall, who expanded the collection of embryos and fetuses, while creating dedicated scientific enterprises at Johns Hopkins University in Baltimore. Mall was known for his keen sense of observation—what his German mentors called *Raumsinn* (literally, *space sense*)—and this must have allowed him to see the value in rigorous comparisons of normal and abnormal specimens to understand human development. By 1913 Mall had eight hundred specimens; by 1917, he had two thousand; today, the Carnegie Institution for Science's collection of human embryos, numbering some ten thousand, is the

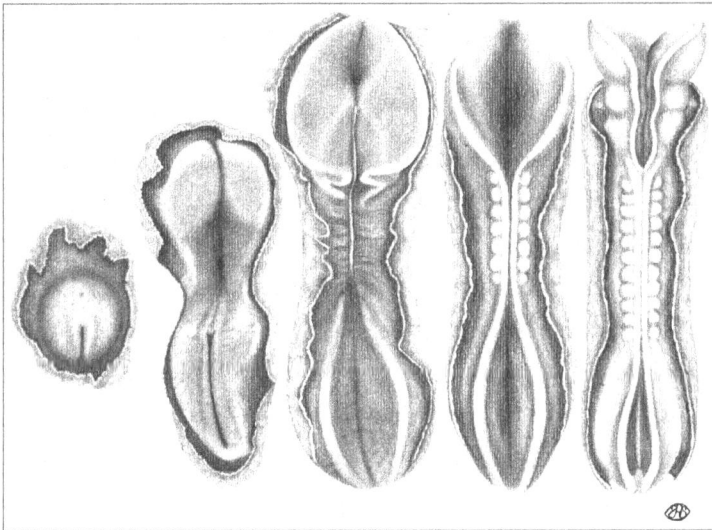

FIGURE 27. Human embryos during early development, spanning from approximately Day 14 (*left*) to Day 28 (*right*). As in the chick, the emergence of the primitive streak (slit on the far-left embryo) indicates the most posterior position of the embryo and presages its anteroposterior axis. The anterior position is demarcated by the cells that will give rise to the brain (the large lobed structure). The cysts organized bilaterally are somites. Images drawn from photographs of embryos in the Carnegie collection.

standard reference for any study of human development. Following His's example, Mall used the names of these specimens to pay tribute to the doctors and scientists who supplied them.

The information to be gleaned from the very early embryos in the Carnegie collection is remarkable in many ways. The collection contains multiple examples of the period when the embryo is between one and three millimeters in length—around when gastrulation begins and the primitive streak starts to plough its furrow. These reveal that around the time of gastrulation, the embryo is not only very small but also buried within a nest of protective membranes, like a walnut in its shell. Also remarkable is the fact that Mall was able to collect them at all. Many women do not realize they're pregnant until the seventh or eighth week of gestation, and many miscarriages at this stage pass unnoticed, mistaken for a late period. For Mall and his suppliers to collect pristine embryos at the age of two, three, or four weeks in the 1910s is remarkable, though the ethics behind how they did it are likely to raise contemporary eyebrows.

Most of the specimens relied on accidents and were derived from miscarriages or abortions with the best specimens coming from gynecological operations; all were obtained without informed consent. Just a few decades later, an obstetrician named John Rock was running a program that would cross further boundaries, as women seeking voluntary hysterectomies at Boston's Free Hospital for Women were asked to have sex with their husbands at a specified time in advance of the operation. Rock was essentially getting embryos to order. This may have allowed him to pinpoint the age of embryos to the day, but a contemporary researcher attempting to duplicate such methods would likely end up facing prosecution for such an invasive procedure.

As ethically murky as their collection may have been, the specimens in the Carnegie collection reveal much about processes that are otherwise obscured by the 14-day rule. Many of the embryos have been serially photographed, sectioned, and reconstructed to provide insights into the emergence, growth, and organization of our body plan and internal organs. Because of the Carnegie collection, we know our heart and other muscles emerge from the mesoderm, the lungs and gut from the endoderm, and the brain from the ectoderm. From reconstructions of development based on the collection's specimens, we can intuit that our cells sculpt themselves from an amorphous mass into the discernible outline of a human body by four weeks after fertilization. The collection has even allowed us to date the appearance of the feature of our bodies that most of us believe truly defines us as humans: our brain.

AN ESSENTIAL DIFFERENCE

Our brains, the wondrous organ that conjures our personalities, hopes, and dreams, contain about eighty-five billion neurons, all generated from a few thousand precursor cells during gastrulation.

Multiplying these initial cells occupies an enormous amount of embryonic activity—at its peak, the embryo creates ten thousand neurons per minute. At every one of these divisions, the machinery inside each cell has to copy, in the correct order, every one of the six billion nucleotides in the human genome so that the next generation of cells has its own copy. Unsurprisingly, mistakes happen. Errors in the copying and recopying of the genome create mutations, and mutations can lead to cellular chaos.

Though our most common reference points for genetic mutations are carcinogenic phenomena in our environments, such as ultraviolet rays or the toxins in cigarette smoke, much can go awry in a developing embryo as well. Harvard neurobiologist and physician Christopher Walsh, while studying the origin and molecular basis of neurological diseases, grew interested in hemimegalencephaly (HMG), a congenital condition in which localized enlargements of the brain cause epilepsy, often quite severe. Because people with HMG exhibit the defect at birth, he decided to look for a genetic mutation.

He and his lab group found that AKT3, a gene coding for a protein that makes cells grow, was mutated in people with HMG. They noticed that in HMG patients, this tool is hyperactive in clusters of neurons, which explains why individuals sometimes exhibit enlargement only in some regions of the brain. Walsh then examined the DNA of people with HMG and found something curious: though the mutation was present in diseased brain cells, it was not present in all brain cells or in cells from other tissues, like the blood. How could only some of a person's cells feature a genetic disease that should have been inherited by the whole organism? Though the idea was floated that people with HMG might be chimeras, mixing cells from different zygotes with different genomes, subsequent research revealed they were something else entirely: genetic mosaics.[10]

You probably know people with patches of skin of different pigments or a streak of hair of a naturally different color. They are likely to be mosaics. At some point during development, as their cells were dividing and multiplying, a mutation occurred while the genome was being copied by a cell. Mutations seem to be an occupational hazard of cells having to copy so much, so fast, during development. In people with HMG, a copying error like this must have occurred during development and then been passed on to all of that cell's descendant cells but not shared among the rest of the individual's cells generally. Some of these cells ended up in the brain.

Errors happen, and they happen at a relatively constant rate. Starting with this observation, Walsh and his lab found something more surprising: even in cells that are not dividing, errors happen, causing additional mutations to accumulate over time. He estimates that every neuron in an adult human brain carries over fifteen hundred mutations.

Not all mutations arise during development. During our lifetimes, some mutations occur as a result of cells aging, while others arise when cells are subjected to damage that cannot be repaired. Different agents of damage leave a signature that can be read in the DNA. So, for example, smoking changes pairs of the nucleotides CG to AT, and exposure to sunlight changes pairs of CC to TT. The signature seen in aging cells is different: it changes XC to TG, where X is any of the four nucleotides C, G, A, or T. Mutations due to cell aging happen spontaneously, about twenty to fifty times per year in any adult cell, following some internal clock that starts ticking when the zygote first divides into two cells. Most of the time, the mutations have no effect on how cells work, because, as we have seen, only a small fraction of the genome, about 2 percent, codes for tools. But sometimes, in the game of chance, a mutation strikes in a protein-coding sequence, altering the genetic function of that bit of DNA. Suddenly, the cell's toolbox is changed—a tool is removed, altered, or broken. This is what has happened with the AKT3 gene in people with HMG.

Mutations due to cell aging are particularly pertinent to the functioning of the human brain because very few new neurons are created after birth. Walsh and his lab group hypothesize that not only diseases of the brain but also some of the cognitive disabilities

associated with getting older might be due to accumulated mutations like this.[11]

Earlier, when we considered how scientists have linked various
genetic mutations to diseases, I suggested that mutations in a gene
often don't have much to tell us about normal development. And
this is true; mutations in eggs and sperm—the sorts of mutations
that organisms inherit from their parents—don't. However, the accumulated mutations that result from aging are a different story.
These kinds of mutations *do* tell us about development, particularly
since the descendant cells in a lineage are effectively "barcoded" by
the mutations they carry.

Though simplistic gene-centered narratives can seem to imply
the existence of a single genome inherited from a crossing of one
sperm and one egg, genetic mosaicism reveals that this code is always subject to change. Right from our very first cell division, mu-

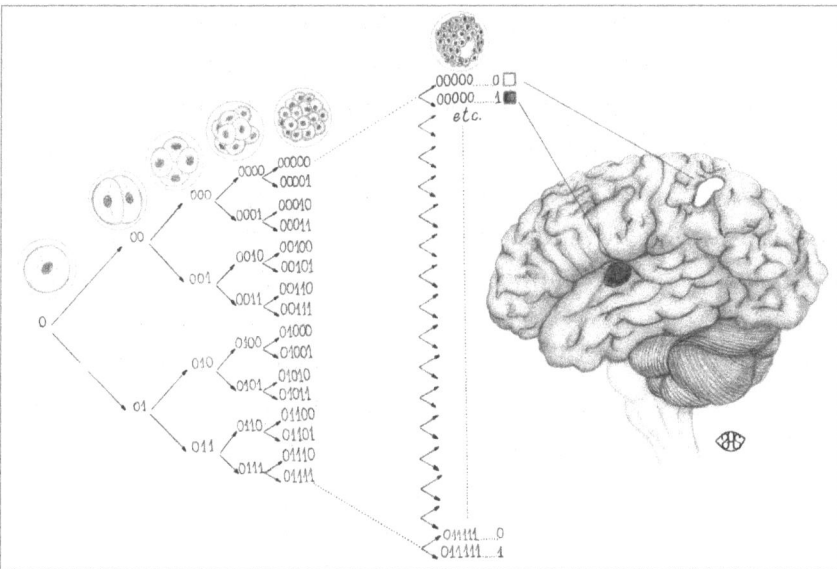

FIGURE 28. Example of cellular barcoding in the human brain. From the first
cell division, mutations happen in every cell. As the mutations are random
and different, this means that over development cells become "genetically
barcoded." They also generate clones that can be related to those barcodes.
Shown: cells in two clones—white and black—each derived from a single cell,
have accumulated different mutations. In some instances this can cause disease.

tations occur, so that each of the two offspring of the first zygote has unique mutations, about five or six each, making their genomes different from one another. Then when those two cells divide, errors continue, so the offspring cells have the parent mutations as well as their own unique mutations. And it continues down through the generations of cells. From what we can tell, during the genesis of the embryo, mutations mount even more rapidly than in adult cells. And these rapid mutations are happening at the time when cells are finding their fates in the body plan and specializing in terms of their form and function. Perhaps most importantly, these mutations and their lineages offer scientists a new way to turn back the clock and reconstruct the earliest processes of development.[12]

It is now possible to use mutation barcodes to trace organs and tissues back through time to common ancestor cells to get a clue as to their origin. Because each cell adds new mutations to the ones it has inherited from its parent, these barcodes can be used to document the life story of every cell in the body and trace its ancestry backward not just to gastrulation but to one of the two first cells. As Walsh would say, give me a piece of Einstein's brain, and I can tell you what his embryo looked like.

Indeed, by the time gastrulation starts, every cell in the embryo is already genetically different. When we trace mutation lineages in the barcodes, we also find that cells in a wide range of organs and tissues share a mutation, while at the same time cells in the same organ or tissue don't always (as in people with localized HMG).

Alongside everything else, this offers further proof of the cell's predominance over the gene. Though mutations in each cell's DNA can tell us where it came from, the genome only allows us to look backward, not forward. The dance of gastrulation jumbles everything up, from cells' identities to their positions and contributions to tissues and organs. All the while, their DNA is steadily mutating and changing. Cells end up becoming what they are and in turn make us who we are, not through some foreordained code but through their ability to interpret where they sit within the community of cells—during embryogenesis and then as they take their place in a particular type of tissue or organ. Every cell is unique because of this history, and this uniqueness is what makes you *you* and me *me*.

We don't have a single DNA sequence; we have billions of them. In fact, on the assumption that there is one mutation for every cell division, we probably have as many mutations as cells—likely more. What we think of as "our DNA" is something like an "average"— but it's an average based on whatever sample of cells has been taken and doesn't come near to representing the full tapestry of our bodies and our brains, as they've been knitted together by billions of cells.

In light of all of this, are we any closer to determining when an individual human being comes into existence? It may be that the construction of the body plan begins at the very first step in the dance of gastrulation, and if you are satisfied with accepting this plan as the defining feature of your individuality, you can settle for the appearance of the primitive streak as your milestone for coming into being. But based on what we've learned from embryos and genetic mosaicism, we can draw another, more bewildering conclusion: during our lifetime there's no single moment when our cells stop building us into something new.

What a piece of work we are. From our very first cell to our very last breath, we are the continuous creation of a vast community of cells—*our* cells. We have learned much about their workings, the manner in which they interact, even the ways that they create not only new cells but new beings. Furthermore, we have marshalled some of this knowledge and now use it in a practical manner to imitate one of the most amazing constructions, a human being. But we do this by lighting the fuse that lies sleeping in the egg and then letting the cells do their bidding to make an organism. One wonders whether we could go beyond the beginning and use cells to re-create, à la carte, bits and pieces of ourselves and maybe even reverse time, as the cloning experiments with animals suggests could be possible. This might sound to you like the stuff of science fiction, but progress in our understanding of the cell over the last few years has revealed that it is not that far-fetched. We can now work with cells in surprising ways to make increasingly more faithful copies of organs and tissues that one day could be used to repair our own after damage or aging. This work is also raising the possibility of making copies of ourselves and, in the process, challenging our ideas about what and who we are.

PART III
THE CELL AND US

The history of a man for the nine months preceding his birth would, probably, be far more interesting, and contain events of greater moment, than all the three score and ten years that follow it.

—SAMUEL TAYLOR COLERIDGE, "NOTES ON SIR THOMAS BROWNE'S 'RELIGIO MEDICI,' 1802"

The primitive streak stage is a vitally important landmark in development because it marks the onset of individuality. Before this stage we have an assemblage of cells that usually gives rise to an individual human being . . . but sometimes gives rise to two individuals—and indeed sometimes gives rise to nothing. . . . In the early embryonic plate . . . the die is not yet cast.

—ANNE MCLAREN, "WHERE TO DRAW THE LINE?"

Engineering is the art of modeling materials we do not wholly understand, into shapes we cannot precisely analyze so as to withstand forces we cannot properly assess, in such a way that the public has no reason to suspect the extent of our ignorance.

—PERHAPS DR. A. R. DYKES, 1976

seven

RENEWAL

WE CAN'T LIVE FOREVER, EVEN THOUGH SOME BELIEVE WE might be able to come close. There are those who argue that some elixir or magical mixture of diet and physical regimen will give our cells the fuel they need to last much longer than they do, defying the usual biology of aging.

Some even think that this elixir lies in the genes. Among the prophets of immortality are Jeff Bezos, founder of Amazon and, more recently, Altos Labs, an enterprise dedicated to finding a method to reprogram cells in an organism such that, just as in the cloning of adult cells, they will rejuvenate, potentially ad infinitum. Indeed, Bezos and a small group of business partners have been luring some of the best minds in science—including Richard Klausner, former director of the US National Cancer Institute—with the goal of delivering eternal youth to us in the next few years. Developing therapies for aging—as if it were a disease rather than an intrinsic part of our biology—is big business, and considering the billions of dollars that have already been funneled into the effort, Altos may succeed where so many have not.

Cells are not designed to live forever, however. Calico, powered by Google, has been working on the magic formula for over ten years without a useful deliverable to date, but the future is always an unwritten book. Though Bezos's project has attracted impressive names and dollar figures, it has little in the way of scientific advancements to show as of yet; it is a bet on hope. Perhaps this is why the research team at Altos has shifted the terms they use to describe their ambitions; now they're seeking not eternal life or even eternal youth but a healthier, longer life. "Live long, die young," as Klausner put it, almost ironically.

Nonetheless, the search for immortality has a long history. Altos's research ambitions hinge on a series of experiments that raised the possibility that some cells are immortal—well, quasi-immortal. If we can learn the secrets of these special cells, we may well find the key to making our own cells immortal too. But, before we get to meet these cells, we need first to come to terms with their limits.

PASSAGES OF LIFE

In the early twentieth century, American embryologist Ross Harrison wanted to study the neuron in an environment where he could explore its organization and activities both in isolation and in combination with other cell types. He was interested in seeing whether a neuron would develop on its own as well as how it would interact with other cells to form functional structures and networks in controlled conditions. In 1907, working with tissue taken from frog embryos, he figured out how to keep neurons alive and functioning in a culture, which allowed him to observe details of how nerve fibers grew and interacted.

It was the first time a scientist had grown cells in a dish, outside a body, and others were inspired by Harrison's example. His research fired the imagination of Alexis Carrel, a French surgeon turned researcher who was working at the Rockefeller Institute (now Rockefeller University) in New York City. Using Harrison's technique, on January 17, 1912, Carrel began to place groups of cells from an embryonic chicken heart in a culture derived from chicken plasma.

Cells in a culture broth like this tend to multiply and, if all goes well, will grow until they fill all the space in the dish within a few days. At this point, you have to *passage* them—that is, to reseed a few cells from the existing culture into a new culture where they can continue to grow freely, not unlike how one would feed and care for a growing sourdough starter. Carrel wanted to see just how long cells could survive being passaged from culture to culture. He thought that, over the long term, it might be possible to passage cells long enough to, eventually, grow organs that could be transplanted to replace failing organs in a body, what we today would call *regenerative medicine*. The dream was on.

Carrel successfully passaged cells from culture broth to culture broth over a month, then over two months, then over a year. Ten years later, the cultured cells had been passaged 1,860 times, and they were still going strong. The *New York World* breathlessly reported that if the same number of cells had grown in a single chicken, it would by then be "big enough . . . to cross the Atlantic in a stride." After Carrel returned to France in 1939, one of his trusted colleagues, Albert Ebeling, continued the work of passaging cells from the culture many more times, only ending the experiment in 1946, two years after Carrel's death, when he was ready to move on to other things. In a journal article published in 1942, Ebeling referenced the myths surrounding the cells' growth, calling them "Dr. Carrel's immortal chick heart." The beguiling message of this three-decade experiment resonated loud and clear: cells can be immortal, if you know how to take care of them.

If, in the right conditions, cells can live forever, doctors' and scientists' next logical step was to "grow" tissues and organs. If these could be kept alive in a culture, they could be held in reserve indefinitely for future use in transplants and other surgical repairs. Just as excitingly, the cultured tissues and organs could be used to do experiments that could not be done on living beings, for either practical or ethical reasons. Carrel's cultured cells had already been used to test drug toxicity—far wilder possibilities awaited if they could solve the technical challenges.

Yet doubts were mounting about the underlying experiment. Science is built not by reputation but by replication. Though Carrel's research carried a veneer of credibility thanks to his 1912 Nobel Prize for his work on surgical techniques, researchers and physicians found it impossible to duplicate his findings. No matter how hard they tried to repeat Carrel's long chain of passages, at some point the cells died in culture. Often, when a researcher reports an amazing result in a biological system and others can't reproduce it, vanity takes over, and the researcher claims that only he or she has the genius and skill to do such experiments. Karl Illmensee did this when he falsely claimed to have cloned mice, as did some preformationists when they pretended that they could observe little humans in egg or sperm with their superior microscopes. In this same vein, Carrel often argued that only he and Ebeling had mastered the tricky techniques required to passage cells over and over, and everyone else was simply not doing it right. For years, this story stuck, and the immortality of cells remained dogma for two generations of scientists.

It finally came crashing down in 1961 when American cell biologist Leonard Hayflick openly challenged Carrel's claims. He'd done many experiments along similar lines, and time and again he found that cells in culture were not only mortal but possessed a life span with clear limits: they could only divide about fifty times before reaching what biologists call *senescence*, from the Latin *senex* for *old age*, and dying. When Hayflick submitted his research to the *Journal of Experimental Medicine*, the editor—none other than Peyton Rous, who won a Nobel Prize for the discovery of the Rous sarcoma virus—rejected the paper for publication, responding, "The largest fact to have come out from tissue culture in the last fifty years is that cells inherently capable of multiplying will do so indefinitely if supplied with the right milieu in vitro." The setback was only temporary, fortunately. Enough other researchers, Nobel Prize winners or not, had also observed the inevitability of cellular senescence, and Hayflick's research finally appeared in the journal *Experimental*

Research.[1] Today, the maximum number of divisions that a cell in culture will undergo is called the *Hayflick limit* in his honor.

There are many hypotheses about what Carrel did with his cultured cells. Charitable interpretations suggest that the cultures were inadvertently contaminated with new cells at every passage, with new embryonic cells added to the culture to "feed" the cells drawn from the old ones. Others can't believe that it could have been an accident. However he did it, we know today that cells cannot live forever—at least, that is, not *normal* cells.

For reasons that are not always perfectly clear, cells are subject to a life cycle. Over the course of a normal life, including in a dish, cells are buffeted by a predictable set of everyday insults—such as radiation from the sun and minerals or toxins in consumed food and water—that cause mutations in DNA. These mutations gradually make the tools and materials the cells need to maintain and repair themselves ineffectual or unavailable. For a while, cells cope with this through a variety of mechanisms, in particular functional redundancies: the existence of backup tools. However, as their inventory of genomic hardware becomes more and more defective, their internal organization gets clogged with defective proteins, and cells deteriorate and eventually die. And this happens not just because the DNA in the nucleus suffers the consequences of these insults; the DNA in the mitochondria—the bacteria-descended powerhouse of eukaryotic cells—also suffers mutations, resulting in decreased energy supply to the cell.

Mutations aside, as a molecule, DNA in and of itself is quite resilient. As noted elsewhere, when a person dies, his or her DNA is still present in each cell and, kept intact and in the right conditions, can survive for thousands of years. DNA doesn't merely survive death; it remains active for a brief period. In mice and zebrafish, for example, transcription continues as late as two days after the death of the organism. The particular group of genes that are being expressed at this moment configure a "signature of death," a specific group of genes operating without any additional activity or feedback from a living cell.[2] Eventually, transcription stops, and the DNA goes into a sort of hibernation and gentle decay.

Though our DNA persists, in some state, even after death, in the end we age because our cells age. It is the functional decay and collapse of our cells that drags us along, all the way to our end.

BREAKING THE PACT

When I said above that cells don't live forever, you'll have seen that I qualified this by saying "*normal* cells." One special type of cell can pull off the trick of immortality under the right conditions: cancer cells.

We have previously explored the Faustian bargain between animal cells and the genome. Cells are allowed to take control of the genome's hardware in order to build and maintain the organism, so long as the cells pass the genome along intact to the next generation through the germ cells: eggs and sperm. This pact works because germ cells behave almost like a bunker, protecting the genome from the insults of everyday life, and the genome allows itself to be used by the cells during development and the functioning of our bodies. This is a fragile arrangement, however, because the genome constantly threatens to seize control of the cell. If it succeeds, cancer takes hold of the organism.

Scientists first learned about the details of this takeover from a sample of cells taken from Henrietta Lacks, the African American woman whose story was told with great humanity by science writer Rebecca Skloot. In 1951, Lacks became ill and was admitted to Johns Hopkins Hospital in Baltimore. There she was diagnosed with cervical cancer. During her treatment, which ultimately failed to save her life, a tissue sample was taken from her cervix and sent to the laboratory for analysis. George Gey, the in-house cancer researcher, received some cells from the sample, as he did for everyone at the hospital who was diagnosed with cervical cancer. With Carrel's immortal cell culture still in good stead, Gey's ambition was to successfully passage human cells in the same way.

Culturing cells from a tumor was of particular interest because the longer he could keep them alive, the longer they could be studied, potentially providing clues to the origin and operation of malignancy.

Though no one had yet been able to repeat Carrel's experiments, one of Gey's lab assistants, Mary Kubicek, set about isolating cells from the sample taken from Henrietta Lacks—some from normal portions of her cervix, some from the tumor. She placed them in a culture in a test tube, labeled it "HeLa," and walked away, fully expecting the cells would survive for no more than two or three days. When she returned to the HeLa culture a few days later, she was astounded to see that the cells taken from the tumor were not dying but instead multiplying at a phenomenal pace, doubling in number every day.

Passaged again and again and tucked safely and securely in a culture, HeLa cells appeared to be truly immortal. They have since become a staple of biomedical research, providing insights into the life of both normal and cancerous cells. Their offspring have been used to test and manufacture vaccines, making fortunes for pharmaceutical companies, and continue to feature in experiments to this day.

HeLa cells tell two stories, one about origins and identity and the other about immortality. In the first story, just as scientists and ethicists found it difficult to identify when a frozen blastocyst becomes an individual and to whom it belongs, HeLa cells raise the same questions. Members of the Lacks family, along with many others, have eloquently argued that because the cells originated from a sample taken from Lacks's body, without her knowledge, they are still *her* cells—and by extension, theirs, because of their shared genetics. At the same time, the cells being used today in labs around the world were never part of Lacks's body, and indeed they are many, many generations removed from the original tissue sample. The only thing they may have in common with her cells is their DNA—and in that case, it is not the same genome her relatives share but a mutated version that went renegade, attacking her body and killing her. How and when we believe that human cells are part of a distinct individual human being and how we use human cells in research are questions that demand consideration that looks beyond DNA for answers.

The second story encompasses what HeLa cells have taught us about biology. Just a few years before Carrel's claims of immortal cells were dethroned, this culture proved that cellular immortality was possible after all. Unlike Carrel, Gey was transparent about the methods his lab used and distributed HeLa cells to any lab that wanted them, along with instructions about how to keep them growing for further research (a free-for-all that the Lacks family found especially troubling when they learned about it).

Yet the way that HeLa cells have evolved reveals what this version of immortality actually means. HeLa cells have continued to grow wherever they have gone, becoming more abnormal over time. Ongoing surveys of HeLa cells at labs around the world show that there is not one single type but a family of HeLa cells that differ in their behavior, the genes they express, and their underlying DNA. These cells are plainly no longer the cells that were in Lacks's body. Her cells, like most people's, had twenty-three pairs of chromosomes; nowadays, HeLa cells have any number of chromosomes, typically ranging between thirty-five and forty-five pairs, and their genomes are mangled in such complex and different ways that scientists can trace the ancestry of every lineage. Many scientists would say that these changes mean the HeLa genomes are no longer technically human. You could never clone Henrietta Lacks from these cells; the tools are broken beyond repair.

Dramatic as they are, these mutations demonstrate how cancer works. The process of multiplication and deviation started in Lacks's cervix, where the genome violated its Faustian bargain with the cell. This probably started with a mutation that unleashed a cascade of events that wreaked havoc in her body. Eventually, mutations frustrated the cell's capacity to repair itself and cooperate fruitfully with its neighbors and promoted rapid proliferation of

cancerous cells, which had a ruinously damaged set of tools and fixtures at their disposal. As the cancerous cells increased in number,
they invaded surrounding territory, destroying tissues and organs.

Yet, because tumors need the body to supply them with nutrients
to fuel their growth, they are also constrained by the rules of biology. They can't kill everything in the body without triggering their
own extinction. So they co-opt their neighbors to selfish ends. The
cancerous cells send a signal to nearby normal cells to get them to
send a set of signals that promote blood vessel creation, or *angiogen-
esis*. Tumors depend on blood vessels—lots of them.

Outside the body, such constraints no longer applied. So it was
that when Kubicek planted Lacks's cells in a culture, where nutrients are constantly available with none of the obligations attached
to existing within a living organism, the cells lost control over their
genome entirely. HeLa cells became mere vehicles for the genome,
in the purest sense espoused by Richard Dawkins. And as the cells
proliferated, the genes were shuffled. Cells that would have died
quickly, lacking access to the tools and fixtures they needed to work
with other cells, now survived. HeLa cells were granted immortality
at a steep price: a set of genes that were no longer useful for much of
anything other than replicating themselves.

In the decades since HeLa cells were isolated, other immortal
cell lines have been obtained in labs. The seeding cells usually come
from a tumor and display similar patterns of mutation cascades and
DNA selfishness. In fact, if you want to make a normal cell immortal, you merely have to fuse it with a tumor cell. And much in the
manner of the creature at the center of the *Alien* films, the cancerous cell will turn the normal cell into a food-seeking partner, until
eventually the cancer destroys its host.

The aberrant behavior of cancerous cells appears to be caused by
mutations in a number of genes with the code for tools central to
cells' self-care. These include tools that cells use to multiply and divide and to communicate during this process, as well as tools used to
check and repair errors in transcription and translation. In the case

of HeLa cells, such mutations differ from one lineage to another. However, one feature is common to all immortal cell lines: the ends of their chromosomes, which scientists call *telomeres*.

The name *telomere* comes from the Greek *telos*, meaning *end*, and refers to the DNA at the end of each chromosome. Telomeres feature a special repeating sequence of nucleotides and do not code for proteins. They serve the same function as bubble wrap around a fragile vase and help prevent damage to the main, gene-coding part of the chromosome. However, telomeres are subject to the same everyday insults as any other piece of DNA. As a result, the length of your telomeres reflects the age of your cells: the older a cell, the shorter its telomeres will be. Nonetheless, there is a tool, an enzyme, that fights against this process; it is called *telomerase*.

The telomeres in HeLa and other immortal cell lines are much longer than those in normal cells. This is because in these cells, *telomerase* is present at higher-than-normal levels. In fact, all cancer cells have high levels of this enzyme. Ever selfish, if it has a chance, the genome, in hijacking the cell, puts itself first.

If you give telomerase to a normal cell growing in a culture, you can get it to keep dividing and multiplying, passage after passage, breaking the Hayflick limit. But this does not mean the cells are any younger. They're able to grow without limit, but they do age. It is a reminder of the Greek myth of Tithonus and Eos, the goddess of dawn. Eos fell in love with Tithonus, mortal son of the king of Troy, and asked Zeus to give him eternal life. Zeus granted this but did not include eternal youth, so Eos found herself having to take care of Tithonus for eternity; in some later versions of the myth, he became a cicada. As the Greek myth suggests, having both long life and health is difficult; Altos should beware.

HeLa cells do not provide us with a road map to our own rejuvenation because these cells are now working entirely at the behest of the genome. But another type of cell in our bodies strives for immortality. These cells rewrite the Faustian pact, amending it to the benefit of the organism.

A NEW YOU

As you read these words, a bitter battle is being waged in your gut. On one side are the by-products of the food you ate at your most recent meal. On the other side are your intestinal cells and their allies, colonies of bacteria that have established a mutually beneficial relationship with your body to help break down nutrients in return for a home in your bowel. For the most part, the cells living in your gut win, but even victory takes a toll. Every day, over a billion of the cells lining your intestines routinely perish. Unless you have one of a set of serious medical conditions, you won't notice their loss. This is because, buried in the lining of your intestine, a collection of special cells is hard at work generating a continuous stream of new recruits so that, about every week, you get an entirely new intestinal lining.

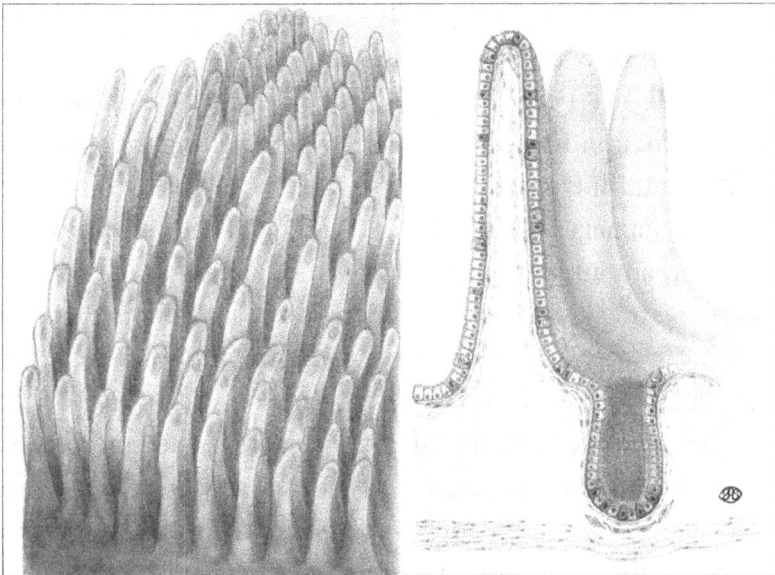

FIGURE 29. The intestine is organized as a carpet, a large epithelial lining, folded into villi (*left*) to increase its surface and facilitate its job of absorbing nutrients from food. *Right:* a cross section of one of the villi displaying the crypt where the intestinal stem cells lie in a niche of surrounding cells. They act as a sort of conveyor belt, working from bottom to top in this image, that produces working cells that do their job and die in the lining of the gut. The different classes of cells are indicated by different shadings.

These special cells are called *stem cells* because they sit at the stem, or root, of a branching diagram of cellular growth and differentiation. They are cells that have some special properties. First, they are a reservoir of the cell types of a particular tissue and can give rise to more than one type of cell: the different cell types that are used to process food. They achieve this because when they divide, rather than making specialized cells, they often make copies of themselves so that the potential to generate the working cells is retained; many live forever, or at least as long as we live. Besides occupying your intestine, they also exist in other parts of your body. In your skin they make new cells to replenish the outermost layers, which slough off or get damaged in the course of daily wear and tear. Thanks to their efforts, you produce an entire new skin every month. Most remarkably, blood stem cells in your bone marrow generate millions of new cells every second. We can thank them for the two billion new red blood cells we use every day to replace those that die in the process of delivering oxygen.

Stem cells age but very slowly. They appear to keep their telomeres and their mitochondria young by controlling their own multiplication and life cycle. They also have a way of fending off the continuous damage we are subject to. Though the cells in your tissues and organs are constantly damaged by toxins, radiation, infection, injury, and other insults, for the most part, your tissues and organs don't stop working. For this we can thank stem cells, busily replacing dead or dying cells with new ones.

With only a few notable exceptions, most tissues and organs in the body have stem cells, *adult stem cells*, as they are called because they are part of your grown-up body. One of those exceptions is the adult heart; the brain was long believed to be another. For much of the twentieth century, scientists maintained that the number of brain cells remained constant as of shortly after birth; if any of your neurons were destroyed during your lifetime, your brain would simply have to work around the loss or operate at a cognitive deficit. However, in the 1990s, following pioneering studies of the 1960s, scientists began to find evidence that challenged this notion. In mice and rats, new neurons are born at high frequency in the parts of their brain used to process smells, called the olfactory bulbs. Considering these species' reliance for survival on a keen sense and memory of smell, this process appeared to make sense. Could a similar process be happening in our brains to help us keep going?

Testing for stem cells in the brain—or any other organ—of a laboratory animal is not difficult. A simple version of this experiment would involve painting the DNA of individual cells with an organic dye or molecule. If the cell does not divide, the label remains in the cell. If it divides, the label is halved at each division, and so the amount of labeled DNA in a cell relative to the starting point provides a measure of the cell's relative age. This technique has been used to determine the rate of *cell turnover*—how quickly a tissue or organ gets rid of old cells and replenishes itself completely with new ones—in different animals. However, doing these experiments in humans is, as with any kind of experiment on us, not ethical and therefore impossible—or at least unadvisable. However, sometimes opportunities arise, and Fred Gage and colleagues saw one in cancer patients who, for therapeutic reasons, had been given a dose of a substance that would incorporate in their DNA. When they examined the brains of these patients after death, they observed that cell divisions had taken place in some regions of the brain, in particular a region of the hippocampus called the *dentate gyrus*, which plays a central role in learning and memory.[3]

This was tantalizing, but would it happen in healthy individuals? From 2005 to 2015, Kirsty Spalding and Jonas Frisén of the Karolinska Institutet in Stockholm ran an imaginative study seeking an answer.[4]

Painting human brain cells in the same way as was done with the cancer patients is too risky. However, Spalding and Frisén realized that a natural experiment, conducted between 1955 and 1963, had done the painting for them already. At the height of the Cold War, when the United States and the Soviet Union were testing nuclear weapons aboveground, the detonation of the bombs had filled the atmosphere with a strange form of carbon called an isotope, carbon-14. When people breathed the air, this isotope flooded into their cells. It remained with them ever after, marking them as having existed at the time of a detonation. In 1963 the testing stopped, and so those people born during the test period had been "pulsed" with the unusual isotope. Chasing the amount of carbon-14 in the tissues and organs of those individuals should provide information about the lives of their cells. This is done based on the properties of the molecule.

The vast majority of the carbon on earth contains six protons and six neutrons. Carbon-14 differs, featuring a less stable structure that includes six protons and eight neutrons. Over time, it slowly transforms itself, or decays, into nitrogen, which has seven protons and seven neutrons, in such a way that precisely every 5,730 years, the amount of carbon-14 in an object is halved. This rate of decay is so regular that archeologists were able to develop the technique of carbon-14 dating to determine the age of ancient organic remains with an accuracy of plus or minus two years. The same technique of carbon-14 dating allowed Spalding and Frisén to estimate the rate of cell turnover in tissues and organs throughout the body, including the brain.

They obtained tissue samples from 120 cadavers of people who had consented to be part of the experiment before their death and applied carbon-14 dating to cells known to turn over regularly. As expected, they found that the intestinal and skin cells in people born during the period when aboveground nuclear tests were being done were much younger than the rest of the individual's cells. Stem cells had done their work. Samples from the cerebral cortex—the layers of folded and ridged gray matter associated with perception and cognition—revealed neurons were full of undiluted carbon-14; none of these cells had divided. However, deep in the hippocampus, they saw signs of carbon-14 dilution in a region called the *dentate gyrus*, as in the case reported by Gage. Much as with rodents and their sense of smell, it makes sense that human cellular renewal would take place in such a place, as it is a center of learning and memory, and such turnover would serve its function.

Spalding and Frisén estimated that a few hundred neurons are born in the dentate gyrus every day during adulthood.

This provocative and intriguing finding has not gone unchallenged. Some recent studies have reasserted the claim that no cells, or very few, are added to our brains after the early years of childhood. There remains much to be uncovered about these processes in the years to come, but something is going on in the mysterious dentate gyrus.

Carbon-14 dating also allowed Spalding and Frisén to survey the age of every other type of tissue and organ in the human body, and they found surprises there as well. By and large, our bodies are younger than we think. In addition to your gut, skin, and blood being regularly replenished, your body builds a brand-new skeleton over the course of each decade. Even your fat cells do not stick around forever, getting renewed about every eight years. They were also able to show that a number of tissues are definitively not renewed after birth, including the lens in your retina and the cells of your heart.

These findings highlight that even an individual human body is not a static structure that slowly decays from birth. Not only can it encompass multiple genomes, but its parts comprise a multitude of ages. Each tissue and organ has its own rhyme and reason; their stem cells run on their own clock, despite having almost the same DNA. The origin of these numbers and differences remains unclear, but one thing is obvious: every year, the cells making up most of the structures in your body change. From year to year and day to day, you are constantly a different individual from the one you were yesterday, made up of different cells and genomes growing ever more different all the time.

PRÊT-À-PORTER

Wondrous though they are, stem cells can be difficult to observe. Unlike tumor cells, which are happy to show up and multiply in a dish, stem cells are shy, happier when cuddled in what is called the *niche*, a cellular environment deep in their tissue of origin, which we have not yet been able to reproduce in the lab.

Outside their natural homes, stem cells quickly abandon their special status, transform themselves into specialized cells, become exhausted, and die. This is why for many years researchers only knew about the existence of stem cells from what they do, inferring their existence from experiments. The first clear indication of the existence of stem cells arose in the 1960s when University of Toronto biologist Ernest McCulloch and physicist James Till wondered how the body managed to supply enough blood to keep us alive given the limited life span of blood cells. Till irradiated a control group of mice, wiping out their blood cells, leading to death, while injecting another group of irradiated mice with bone marrow cells taken from healthy mice. When the irradiated mice survived to enjoy a long and healthy life, Till and McCulloch drew multiple conclusions.[5] For one, the experiments suggested that some cells in bone marrow were capable of establishing themselves in a different body and, despite having a different genome, had the power to renew a new body's supply of blood cells. Out of this research came the development of bone marrow transplants to treat diseases such as leukemia, where a person's bone marrow produces large numbers of malfunctioning white blood cells. Yet the presence and nature of blood stem cells in bone marrow remain a hypothesis, given that no one has yet succeeded in maintaining blood stem cells in a culture.

Nevertheless, researchers have recently started to learn to capture other kinds of stem cells for study in the laboratory, with promising results. During the early 2000s, a group of researchers led by Hans Clevers, then at Utrecht University in the Netherlands, investigated the regenerative capacity of our intestines and, in particular, the genomic tools used by stem cells in the gut. Clevers and his colleagues found that these stem cells in mice had proteins on their surface that allowed them to be identified. The next challenge would be to isolate them in a culture.

A gastroenterologist in the group, Toshiro Sato, decided to rise to the challenge, using the markers associated with them and tin-

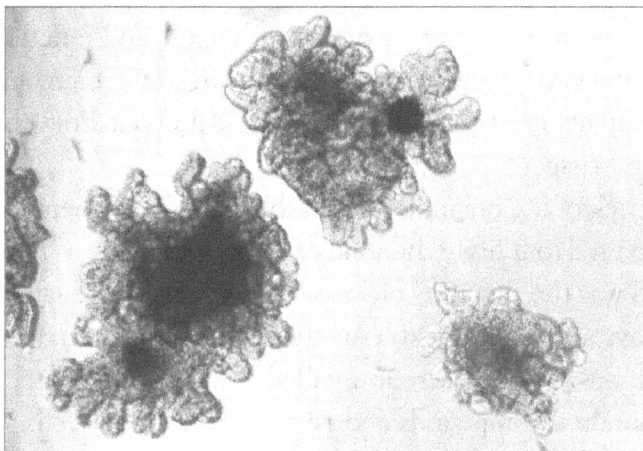

FIGURE 30. A human intestinal organoid, or minigut, generated from adult intestinal stem cells.

kering with the culture conditions to try to replicate the niche in the intestinal lining. Once Sato had isolated them, the stem cells did something surprising. After a few days, they had not only survived and multiplied but also built spiky, hollow spheres of cells out of ribbons of regular intestinal cells, with a few stem cells spaced at regular intervals at their bottom. The cultured stem cells had created a three-dimensional replica of a stretch of gut—a minigut.[6]

After a few more days, this minigut had grown too large for its dish and had to be broken up. Much in the manner that a gardener might use cuttings from one specimen to grow new plants, the pieces of minigut were used to seed new cultures, each of which generated more miniguts, in a process that could be continued, in principle, indefinitely. A cell had become immortal in a culture not by becoming malignant but by reinventing itself.

I have chosen the word *reinventing* rather than *re-creating* because these miniguts were not truly guts. Each was an *organoid*, a simplified, smaller version of the organ, that, unlike a real organ, can be kept going forever. Miniguts don't grow blood vessels, which imposes some limits on their organization and function. For example, they don't develop immune cells, leaving them highly susceptible to death, should the culture become contaminated. And they aren't connected to any other tissues or organs, which means they don't exhibit the usual repertoire of behaviors and functions of an organ in the body. However, because we can study them outside the body, miniguts are teaching us a great deal about how cells build organs and tissues.

The trick of the organoids lies in the alchemy of chemical signals that Sato used to imitate the stem cells' natural niche. One of his ingredients was the signaling protein Wnt, which, as we saw in Chapter 5, plays an important role in the choreography of gastrulation. Another was a gluey secretion called Matrigel, composed of over fifteen hundred compounds and derived from tumor cells. Without Matrigel nothing happened. Its role appears to be the maintenance of the epithelial structure of the stem cells, which is an essential element of their identity. However, even when Wnt was in the company of Matrigel, the experiment did not work every time.

Importantly, stem cells were more likely to thrive and build miniguts when, instead of starting the culture with one cell, Sato started it with two—a stem cell and a normal intestinal cell like the ones that would have been the stem cell's neighbors in the gut. Having two cells talking to each other is a more efficient way to build body parts than relying on one cell alone. In fact, when a stem cell is on its own, the first thing it does is make a copy of itself, and the next thing it does is create an intestinal cell. Then this duo builds the minigut together. To build an intestinal organoid, you need two cells cooperating.

This cooperation is more than chemical. Cells also use information from the geometry of the growing tissue and the forces generated by the masses of cells pushing against each other. My colleague Matthias Lutolf, now at Roche, has used microfabrication and engineering techniques to shape the growth of the miniguts into that of the intestine, turning the sacs spontaneously generated by the stem cells into longitudinal structures with stem cell–bearing crypts that start resembling the intestine in vivo. This new generation of miniguts will expand the applications of the system, particularly clinically, where the promise is starting to become reality.

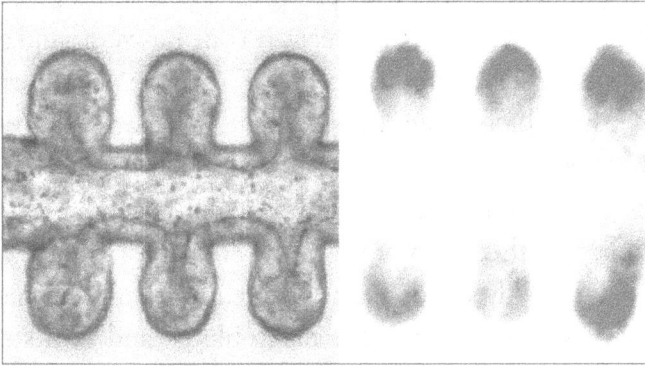

FIGURE 31. Stem cell–derived mouse intestinal organoid engineered to acquire a tubular form, similar to the native structure. *Right:* staining revealing the location of the stem cells at the bottom of the crypts.

Miniguts from adult stem cells are now being used to model diseases and infections and to test drugs in ways that would be expensive and often impossible in animals. In one remarkable experiment reported in 2018 by Nicola Valeri and his colleagues at the Institute for Cancer Research in London, miniguts were cultured from stem cells taken from cancer patients, and the effect of cancer-fighting drugs was monitored in both the miniguts and the patients' intestines. The miniguts and the real guts showed similar responses to the drugs.[7] This suggests organoids could be used successfully to test and screen which drugs, and how much of them, will prove the most beneficial or least harmful to a patient, without risking the patient's health. In an even more striking proof of concept, mouse miniguts have been transplanted into mice, where they have integrated into the intestines, opening up the possibility that one day human organoids might be used to repair damaged tissues and organs.[8]

If stem cells from the intestine can make miniguts, and most tissues and organs have stem cells, it should be possible to grow organoids of most parts of our body. And indeed, over the last few years, various organoids have been grown: minilivers, minilungs, minipancreases, and organoid skin. Much of this work has been done with mouse stem cells, but increasingly human stem cells are being used. In all cases, Wnt signaling and Matrigel appear to be core ingredients in getting the cells to create structures that mimic what happens in the body. Curiously, no one has yet found a way to get blood stem cells to produce blood. The niche deep in the bone marrow hosting blood stem cells is different in ways that we have yet to figure out.

While the various types of adult stem cells each have a defined job in support of the organ or tissue they're part of, another kind of stem cell in the body stands apart. Where intestinal stem cells make intestinal cells, blood stem cells make blood, and skin stem cells make skin, this special kind of stem cell has the ability to give rise to any and all of the cells in your body, including other kinds of stem cells that keep each of your organs going. They're called *embryonic stem cells*. Discovered purely by chance, they offer tantalizing hope to those seeking a magical elixir for cellular youth.

THE ISLAND OF THE EVER YOUNG

In the 1950s, Leroy Stevens, then a newly minted PhD doing research at the Jackson Laboratory in Bar Harbor, Maine, was asked to investigate whether cigarette paper was causing lung and other smoking-related cancers. As was typical, he conducted his study in mice—in this case, a strain referred to as 129. One day he noticed one of his mice had a large swelling in its scrotum. When he sliced open the testicles, he found a strange jumble of different kinds of cells: teeth, hair, and muscle were all mixed up in a lump along with some nondescript cells that he could not immediately identify. Such growths—called *teratoma*, from the Greek *teratos*, meaning monster—had fascinated doctors for centuries.

Teratomas were usually benign tumors that grew in testicles or ovaries, but Stevens was intrigued by their potential for growth. He decided to inject these cells under the skin of other mice to see if they would produce a teratoma in others too. He also injected them in the abdomen of some adult mice. He found that the cells thrived in peritoneal fluid, a plasma that lubricates all the abdominal organs, forming strange growths that, to a trained eye, looked remarkably like the developing ball of cells before they've gone through gastrulation. Intrigued by this similarity, Stevens wondered whether cells from an early embryo would also give rise to a tumor if they were injected into the testes of his strain 129 mice. They did, and Stevens would spend the next two decades observing the behavior and structure of cells harvested from teratomas, which he called *embryo carcinoma cells*. His patient and carefully reported observations were paving the way to that fountain of youth that is the embryonic stem cell.

Stevens's cells, in particular their similarities with embryos, attracted the attention of scientists. How could tumor cells revert to an embryonic stage? One of these scientists was embryologist Beatrice Mintz of the Fox Chase Cancer Center in Philadelphia, who in the 1960s had developed a technique for creating chimeric mice by combining cells extracted from one black mouse and one white mouse blastocyst when each had only eight cells. In 1975, she reported that injecting strain 129 embryo carcinoma cells from a black mouse into the blastocyst of a white mouse produced adults that exhibited skin peppered with white and black patches—a healthy chimera created from tumor cells. Deeper analysis revealed that the strain 129 cells integrated into every tissue in the adult mice except their germ lines—the eggs or sperm, which were infertile. This was intriguing, given that the embryo carcinoma cells came from germ cell tumors, as it suggested that development could reverse a tumor in the sense that cells grew normally and participated in normal development. Surprisingly, this line of research has not been pursued.

The behavior of the embryo carcinoma cells raises the question of whether cells with similar abilities to participate in the development of a mouse exist in normal animals. Matt Kaufman and Martin J. Evans, then at the University of Cambridge, and, working separately, Gail Martin of the University of California, San Francisco, were the first to get results in 1981. When they took cells from very early stages of mouse development, specifically from blastocysts before implantation in the womb, and placed them in conditions that favored growth, small colonies of cells formed that behaved like embryo carcinoma cells. They injected these cells into blastocysts, and just as in Mintz's experiment, they gave rise to chimeric adults. But, unlike the chimeras made with embryo carcinoma cells, these mice were fertile. Martin named them *embryonic stem cells* because they were drawn from the developing animal and, like other stem cells, seemed to have the potential to generate cell types other than their own.

If mice had embryonic stem cells, humans ought to as well. Their blastocysts are very similar, and their cells share the same tools. The hunt was on, but it took nearly two decades, until 1998, before human versions of these magical cells would be derived. This was achieved using cultured blastocysts made possible through IVF. The American scientists who achieved this, John Gearhart and James Thomson, managed to grow cells that, in culture, had many of the properties associated with the mouse cells, but there was a catch. How could they prove their pluripotency, their capacity to give rise to the three germ layers of the embryo: the ectoderm, mesoderm, and endoderm? How could they show that they were true embryonic stem cells? The standard way to test for pluripotency is to mix them with the cells of an early embryo to form chimeras, but as with stem cells in the brain, you cannot experiment by injecting putative embryonic stem cells into a human blastocyst and seeing what they do! So, they bypassed this experiment by injecting the putative human embryonic stem cells into an immunocompromised mouse. If they were bona fide stem cells, they should grow, and if they were pluripotent, the cells would form tumors exhibiting all sorts of cell types, as mouse embryo carcinoma and embryonic stem cells do. The putative human stem cells did just that, and to this date, this is the assay used to test pluripotency of human embryonic stem cells.

The discovery of human embryonic stem cells ushered in a new era of medical innovation. If researchers could learn how to direct the embryonic stem cells to generate groups of tissues, perhaps they

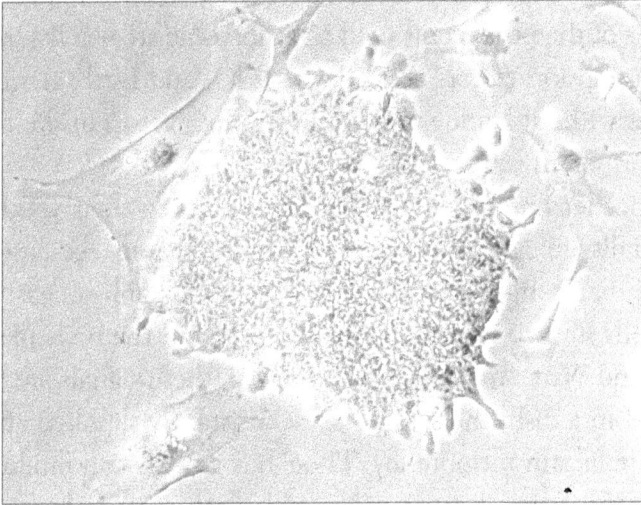

FIGURE 32. An embryonic stem cell colony with differentiating cells at the edge.

could grow nerve cells to repair injured spines, allowing someone who had been paralyzed to use their limbs again. They just as well could be induced to grow heart cells that could be injected into the scar tissue that forms from a heart attack, helping to heal the heart. They might even be persuaded to grow brain cells that could replace the handful of dysfunctional cells that cause the symptoms of Parkinson's disease. The list of applications is practically endless. If a proverbial fountain of youth is indeed out there, it may very well exist in the early days of development, in the undifferentiated cells that carry the power to specialize into any cell type.

Though research on these cells is still in a nascent state, we have begun to uncover how embryonic stem cells operate. For one thing, we have learned that these cells turn to a very small toolkit—three proteins, four transcription factors—to maintain their own youthful state of suspended animation. One of these in particular is critical. Discovered by Ian Chambers and Austin Smith in Edinburgh and Shinya Yamanaka in Kyoto, it was named Nanog, for an island from Gaelic mythology called Tír na nÓg, where time was said to stand still. Nanog is essentially a molecular wrench, used to implement the activities of the other tools used by embryonic stem cells, including the transcription factors Sox2, Oct4, Klf4, and Esrrb. It would appear that without Nanog these transcription factors in the cells lose their pluripotent status quickly.

In another important aspect of maintaining their youthful potency, embryonic stem cells saturate themselves in a cocktail of signals, familiar from the moments in the choreography of gastrulation when cells start to take their final positions in the body plan: FGF, Nodal, and Wnt. In a developing animal, these signals are fleeting. Cultured in a dish, these signals are constantly supplied and allow cells to maintain totipotency. Though scientists can modulate the levels of these signals from the outside with chemicals, embryonic stem cells secrete the signals themselves, seeming to sense that they aren't in an environment that can support gastrulation and a fully developing embryo.

As we have seen repeatedly, cells use signals—the interactions and conversations between cells—to make decisions about their fates and identities. It is the signals that act on the circuits that have been preselected to maintain the cells at the top of the Waddington landscape or to guide them through the hills and valleys that lie ahead. But then, maybe this knowledge could be used to drag the cells up the landscape, from their terminal valleys to the summit.

BACK TO THE FUTURE

After seeing the effect of transcription factors on embryonic cells, Shinya Yamanaka had the audacious idea that perhaps he could take the process into his own hands. The right cocktail of molecules might very well turn an adult cell into an embryonic stem cell. There were antecedents in John Gurdon's cloning experiments, as well as in Dolly the sheep, suggesting that this could happen, but the experiment remained something of a gamble.

Yamanaka started by collecting the names of all the genes expressed in pluripotent cells derived from mice, with a particular focus on those exclusively expressed in the early stages of development. He ended up with a list of twenty-four. Then he and his team placed all of these genes into a *fibroblast*, a cell used in repairs of the body. This created a chemical environment that forced cells to make proteins that were not normally part of their being. Sure enough, after a few days, a small number of the fibroblasts had turned into pluripotent cells. One or more of these twenty-four genes contained a code, a sort of file recovery "time machine" app, that restored cells back to an earlier stage of development, at the summit of the Waddington landscape, before differentiation.[9]

Yamanaka had proven that the transformation was possible; he now needed to narrow the recipe down to find the smallest number of genes necessary for performing this time-traveling feat. It ended up being a combination of four: *Sox2*, *Klf4*, *Oct4*, and *Myc*. All four genes code for transcription factors, and three are key to maintaining pluripotency. Surprisingly, Nanog is not part of the cocktail of proteins. However, Nanog is eventually turned on by the combined activity of the factors encoded by these genes, and Nanog's activation is the signal that the cells have arrived at their destination.

Yamanaka used the four proteins to reprogram mouse cells and then placed the resulting pluripotent cells into an early mouse embryo. The cells integrated into the embryo, helping to build a healthy, fertile mouse that subsequently had healthy, fertile offspring. This proved that the four factors reset the cells to the start of development, allowing them to give rise to any organ or tissue. Yamanaka called the reprogrammed cells *induced pluripotent stem cells*. Later, he and his colleagues repeated the experiment in human fibroblasts, weeding out human genes, and got the same combination of proteins.[10]

The path is different, but the outcome is the same, as in the case of Dolly or Gurdon's frogs. Intriguing as Yamanaka's results may be to the quest for a cellular fountain of youth, the precise relationship between Yamanaka's factors and those that reprogram adult cells in oocytes is not yet clear. Furthermore, Yamanaka's procedure is not very efficient. Tens of thousands of cells receive the quartet of genes, but only one in one thousand is actually reprogrammed by the process. Whether the Yamanaka factors actually work appears to depend on the state of the cell.

Even so, cellular reprogramming helps to lengthen telomeres and improve the efficiency of the mitochondrial powerhouse. The cells *look* younger; Yamanaka's process appears to renegotiate the Faustian pact between cells and genomes. The promise inherent in these findings led in large part to the establishment of Altos, where researchers are testing variations of the Yamanaka cocktail to see if they might one day do their reprogramming trick inside a human.

We shall see. There is no question that something interesting is going on in the cells during reprogramming, and although many scientists are focused on the genes regulated by the Yamanaka cocktail, the most important action is clearly taking place in the cells as whole entities. Whatever the elements of that gene cocktail are doing with DNA, they might also be interfering with other proteins in the cell and rousing elements like mTOR, the control center of cellular activity, into action. It is likely that any sort of breakthrough in this area remains some years in the future, with many surprises along the way.

Nevertheless, induced pluripotent stem cells and organoids have provided a window into how cells use genes to build tissues, organs, and organisms. Importantly they represent a tool for regenerative medicine. For all practical purposes, these induced pluripotent stem cells (often referred to with the acronym iPSC) are equivalent to embryonic stem cells and can be used in the engineering of organs and tissues. Their advantage resides in that they are not derived from blastocysts and therefore bypass the ethical difficulties associated with the use of embryos. From this perspective, their value is immense.

MADE IN PIECES

Research into the capacity of embryonic stem cells to create other types of cells continued to uncover remarkable findings. In 2012 Yoshiki Sasai, a scientist extraordinaire with a keen interest in the development of the nervous system, began to tinker with mouse embryonic stem cells in the hope of learning how the mammalian brain develops. One day, as he checked cells in an experiment being performed by one of his colleagues, he came across quite a spectacle: an optical cup, or embryonic eye, had emerged in one of the dishes, along with some pieces of brain.[11] Eyes are complicated structures, with many different kinds of cells arranged in specific geometries and densities. It was therefore quite a shock to see them emerging in a dish. He tried the same culture and conditions with some human embryonic stem cells and again obtained an eyelike structure and associated brain tissues, this time in an even bigger, three-dimensional version.[12] The cells seemed to know their species of origin and scaled their architecture accordingly!

Where previous scientists had discovered miniguts, a new crop of researchers now unveiled the minibrain. A year after Sasai's discovery, a duo of researchers at the Institute of Molecular Biotechnology in Vienna, Madeline Lancaster and Jürgen Knoblich, unveiled the first so-called cerebral organoid.[13] The small, layered assembly of cells, a mix of neural stem cells, neurons, and other brain cells, had been grown in a culture from induced pluripotent stem cells and had an uncanny resemblance to the folds and ridges of the human cortex. It didn't work like a brain—at least not in any sense we'd commonly recognize—because it wasn't connected to a body. Lacking sensory organs feeding it information to react to and learn from, without feedback loops from the heart or lungs to regulate the release of hormones, and possessing no muscles to control, it wasn't much of a brain at all. However, the minibrain was not without its charms, offering a means of studying the effects of disease on human brains.

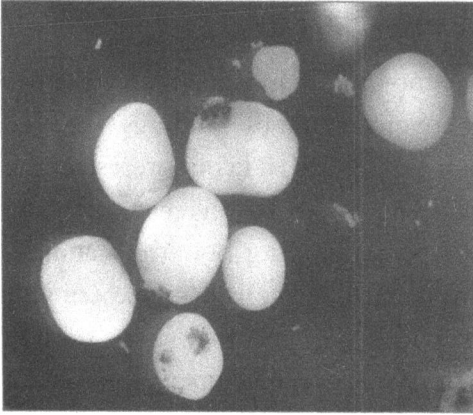

FIGURE 33. A group of minibrains in a dish grown from human embryonic stem cells. The dark spots are associated with pigmented epithelia from eyelike structures.

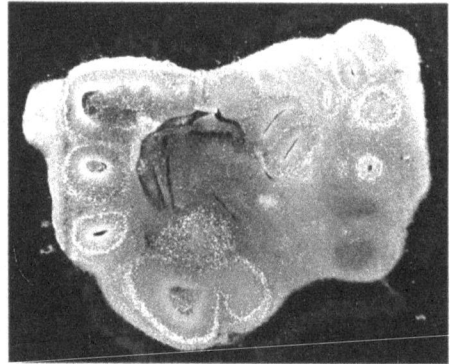

FIGURE 34. Cross section of a human minibrain showing layered structures of neurons that resemble their organization in the cortex.

The cerebral organoid model was put to a major test after 2015, when doctors began to see an unusually high number of babies being born with microcephaly—small heads and a large loss of brain tissue. The vast majority of the cases were traced to infection of the mother and developing child with the Zika virus, which is transmitted to humans by mosquitoes. Health authorities responded swiftly so that the spread of the virus was controlled, but Zika virus was not eradicated, and no treatments for infection exist. If there were another outbreak, it would be useful to understand why brains failed to develop properly in some, but not all, babies born to mothers infected with the Zika virus during pregnancy.

Experiments on human brains are usually made impossible by ethical and practical issues. Naturally, experimenting with Zika virus infection on living human brains is amoral in the extreme. And unlike with other organs and tissues, using other animals to do the experiment isn't an option. This is because the cortex in the human brain is larger, relative to other parts of the brain, than it is in any other animal. The way our neurons and neural circuits process information similarly appears unique to humans, involving a very large expansion in the number of progenitor cells to brain cells early in development, before they've specialized. The combination of numbers and circuitry seems to bestow on us cognitive powers not matched by any other species but also makes human brains incomparable for scientific purposes. To study how the human brain develops and what stops it from developing and operating normally, you simply need to use human or at least primate tissue. Fortunately, minibrains allow us to do this ethically for the first time.

By infecting minibrains with Zika virus at different stages of the structures' growth in culture, Guo-li Ming and Hongjun Song, then based at Johns Hopkins University, were able to demonstrate that the virus had a strong affinity for attacking progenitor cells of the brain. When the virus infected these cells, it curtailed their expansion; the meager progeny they generated often died too, so that the minibrain was much smaller than uninfected minibrains grown from similar seed cells.[14] This is not a cure for microcephaly but instead a step toward understanding the causes of the disease.

Minibrains could also provide clues to the origins and mechanisms of—and potential cures for—other brain-related conditions, including Alzheimer's disease and autism. Though both have received a great deal of attention over the past decade related to possible associations with genes, in at least 20 percent of people with Alzheimer's, genes don't seem to play much of a role. Only a couple hundred people around the world are known to have a form of Alzheimer's caused directly by genes, and no more than half of people who develop the disease have a form of the *APOE* gene that increases their susceptibility to Alzheimer's—and this situation, a probability associated with a mutation, is for the strongest associations of the disease with a gene. Similarly, when you dig into the studies saying autism is genetic, you find that numerous genes may be associated with the risk of developing autism. Clearly, not all brain cells are using the genes in the same way. There must be something else going on, and minibrains offer a way of learning more with a focus on cells.

By combining organoids grown by embryonic stem cells with Yamanaka's reprogramming technique, it becomes possible for scientists to create personalized copies of embryonic cells that, in principle, could generate a matrix of cells for a specific tissue or organ. Take some cells from a person with a neurological disorder, put them in the cocktail that nudges them to become induced pluripotent stem cells, and then put those in the cocktail that nudges them to build minibrains. You can then do unfettered experiments on these personalized organoids, first to probe the causes and consequences of the disease in an individual and, second, to try out which drugs are most effective in ameliorating the disease in them.

Though scientists still have a long way to go, many research programs are using organoids to tease out the origins of cellular dysfunction. Along with personalized brainlike structures, induced pluripotent stem cells can produce muscles, pancreases, and intestines. Each of these lines of research opens up exciting possibilities for a truly regenerative medicine, where a whole replacement organ might be grown by cells recruited from your own body.

Still, our understanding of how cells interact with each other is incomplete. There remains something unsettling about these disembodied structures appearing in a dish, in pieces. Are we just a coalition of organs and tissues rather than the whole that we see and enjoy?

For many years, I worked with the fruit fly, *Drosophila melanogaster*, trying to understand this very relationship between genes, cells, and tissues from a different angle. I experienced a similar queasiness about how such structures are made. I knew that early in their development, at about the time of gastrulation, small groups of cells, called *imaginal discs*, are set aside in specific regions of the fly embryo. When the maggot hatches out of the egg, these groups of cells grow, fed alongside the active cells as the maggot wriggles around the environment, eating and growing. At a certain moment later in development, the maggot forms the pupa or chrysalid, the state of being encased in a cocoon- or nest-like covering. Inside the pupa, something absolutely amazing happens. The maggot cells die, and the now not-so-small clusters of cells that had been set aside at gastrulation come together and assemble the adult insect, like Legos. A closer look would reveal that each group of cells represents a distinct part of the body—a cluster for each leg, a cluster for each eye, a cluster for each wing, and so on. Not unlike a minibrain or miniliver emerging in a dish, each of these components of the adult body has developed independently from the others, but they all fit together—except that the mini-organs don't come together but rather develop together in the embryo.

It may seem like a strange way of building a body, but the very existence of organoids hints that humans are assembled in a similar fashion. If we can obtain guts, brains, and livers in a dish, independently of each other, perhaps we're not so different from flies, and our assorted parts will come together more easily than we might think.

The ability to grow organs, in particular minibrains, from induced pluripotent stem cells has given rise to the notion that they can be used to do something like cloning. Take one of your skin cells, this view suggests, turn it into an induced pluripotent stem cell, then grow a minibrain, a minigut, a miniliver, a minipancreas, and, voilà, one might assemble it all together into a body, like a fly, and you'll have a mini version of you.

However, remember that every cell in your body is different, unique, and this includes its exact configuration of DNA. While cells are undergoing Yamanaka's reprogramming protocol, they are also becoming different. In the end, a minibrain will no more be your brain than the cloned Copy Cat was Rainbow, her donor parent. Significantly, the minibrains that made headlines around the world are exceptional specimens, operating very differently than a brain in a body, whose connections are the work of a lifetime, something that is missing in the structures grown in the lab.

THE ILLUSION OF LIFE

If you happen to be in Switzerland, it's worth taking a jaunt to the lakeside town of Neuchâtel. There, on the stage of a small amphitheater in the Musée d'Art et d'Histoire, you'll find three small, childlike figures, about two feet tall, dressed in eighteenth-century costumes—a musician, a draftsman, and a scribe—which are the handiwork of the master watchmaker Pierre Jaquet-Droz. If you are lucky to be there on the first Sunday of the month, you'll have the delight of seeing them spring into action.

The musician might set things in motion, delicately playing her miniature harpsichord by pressing tiny keys with her tiny fingers. Next, the draftsman might raise his pencil to sketch out one of his four pictures—King Louis XV, an aristocratic couple, a dog, or Cupid driving a chariot—occasionally blowing away any graphite dust. The scribe is particularly mesmerizing. He begins by dipping his quill in an inkwell filled with real ink and blotting any excess ink on the paper placed on his desk. He then starts to write. He can be programmed to write any text up to forty characters long, his eyes following the calligraphy as each letter is formed. Some consider the scribe to be the first programmable computer, six thousand pieces harmoniously assembled to work in tandem, including the gears in his back where the message he writes is set.

Around 1770, when Jaquet-Droz was building these contraptions called *automata*, a lively debate was underway about what comprises an organism. Our understanding of the physics of the planets was a mere century old, and mechanical models of the solar system, called *orreries*, were in vogue. These models mimicked the rotation and orbit of each planet, using cogs and wheels to keep them clinging and clanging in perpetual motion. If the universe was a piece of clockwork, why not also living systems? Jaquet-Droz's automata were an extension of this, an attempt to build life, or the illusion of it, with cogs, plates, and wires. However, even the most masterful automata lack two fundamental qualities that separate organisms from machines: they don't reproduce, and they don't heal.

If Jaquet-Droz were building automata today, he would surely have used cells rather than metal. In place of cogs, plates, and wires, our modern simulacra of life, the organoids, are built with cells. But whereas Jaquet-Droz controlled every aspect of his creations, programming how and when they move, in organoids the cells are in charge. When we make an organoid, we can only put cells in different conditions and watch what they do. We can move groups of cells from one potion to another and try to intuit which signals, culture conditions, and physics will tempt them to become a mini-gut versus a miniliver versus a minibrain. In some *protocols*, as these recipes are called, there is an element of luck in what the cells do. Photographs of organoids are stunning, but we should admit that we do not yet understand these cells' handiwork.

Further, when making a specific organoid, particularly from embryonic stem cells, scientists necessarily take shortcuts. By pushing cells in specific directions, we disregard that all organs and tissues are born in an embryo and that their seed cells exchange signals with each other in ways that affect their development. The heart needs the endoderm and the gut needs the mesoderm to develop properly. It's possible to derive a pure population of one specific cell type, but when this is done, things don't progress as they should. The resulting organs are incomplete and remain in a fetal state, incapable of performing as they should and would in an adult. Organoids don't have the rich and diverse confederacy of cells that embryos do.

It was this idea that inspired me in 2003 to turn my attention away from fruit flies, which I had been working with for fifteen years, to embryonic stem cells. Organoids were intriguing, but would it be possible to create a mini early embryo, the whole thing, with many, perhaps all, of the seeds for the different tissues and organs interacting as they usually do, from embryonic stem cells? How much could cells do together in a dish?

The notion of a mini embryo in a dish sounded impossible at the time. But others had shown that mouse embryonic stem cells could be placed in a foreign mouse blastocyst, and the stem cells would pitch in with gastrulation and build an embryo alongside the descendants of the original zygote. If they could assimilate into their new neighborhood, there must surely be some way to make embryonic stem cells feel at home like that in culture and then let them dance.

– eight –

THE EMBRYO REDUX

THE BLACKBOARD HANGING IN THE CALIFORNIA INSTITUTE of Technology office of famed physicist Richard Feynman, like that of many a scientist, was covered in chalk notes. Naturally, many of these notes were equations. Some were homework assignments. Others were axioms, meant to guide his work and the work of his students. One in particular held pride of place, and because it was on the board at the time of Feynman's death, it has attained legendary status among scientists. It said, "What I cannot create, I do not understand."

Feynman studied the forces of nature, but I take this chalked axiom to be a singular intellectual challenge to modern biologists. If genes are truly the book of life, the operating manual that helps you develop from a single cell, we should be able to use them to build our own biological systems. Indeed, over the past two decades, an entire industry has arisen to engineer gene-based circuits to perform tasks for us, mostly by using the genes in bacteria, as is the case with CRISPR, or to produce proteins of commercial or biomedical value, for example, insulin. But as we have seen, when it comes to engineering tissues and organs, the genome is insufficient unless you've got the right types of cells in the right sorts of conditions, and even then, most of the time we're left clueless. Biology has essentially put a kink in Feynman's axiom, declaring, "What I can create, I still do not, yet, understand."

To understand our identity and origins, we have to learn how cells make us. For all that we have learned, if we still cannot coax cells into building an embryo under our observation, what have we truly understood?

PIECES OF A PUZZLE

If anyone tries to tell you that a fly is simpler than a human, they don't know much about biology. Without a doubt, the *Drosophila* fruit flies that Thomas Hunt Morgan and scores of researchers use to learn about genes are unlikely to compose *La Traviata* or invent the iPhone, but like most ordinary humans, they sleep, get depressed when they're alone, court each other with song and dance, and even count.[1] They also come down with diseases that often mimic our own. About 75 percent of the genes associated with human disease are also present in flies, which allows us to use mutations to duplicate particular human diseases and see how they affect the flies. For example, when *Drosophila* genes are given the same mutations that produce Parkinson's disease in humans, the flies develop neuromuscular atrophies and their consequences, including tremors, reminiscent of those experienced by people with the disease. For this reason, *Drosophila* offers a useful model to study the relationship between genetic dysfunction and cellular dysfunction.

Useful as this research may be in highlighting how cells use genes to build and maintain the body, flies and other insects are made in a remarkable way that, at first sight, differs from how vertebrates are formed. As we saw in the last chapter, each part of a fly's body develops in isolation from the others, as an imaginal disc. Our ability to grow specific organoids independently of each other might suggest that, in a way, organs are like sophisticated imaginal discs. Despite many similarities, the strategies are different. We are interested in those strategies.

During the 1990s, my research team at the University of Cambridge was focused on how *Drosophila* cells make a fly. We were especially interested in how cells communicate with each other in the process of building specific body parts, including how they manage to allocate and pattern the right number of cells to each imaginal disc, transforming a mass of cells into a wing or a leg. We spent a lot of time learning how the fly cells use the Wnt and Notch signaling pathways to become different from each other and organize first the embryo and, in particular, the wing. This does not mean we were successful in answering such questions, but we contributed to progress. At the time, the science of genetics was very much at the fore, so we dedicated much of our effort to associating genes with functions—well, as you know by now, mutations with genes and these with functions—like everybody else at the time. For the historical reasons discussed in Chapter 1, the fruit fly had become the workhorse for this job, but this is an organism in a hurry; development is fast, and when things are happening so quickly, with body parts being built in minutes and hours, it is difficult to pose your questions to the cells. Instead, you can only do the standard science of genetics: knock out or mutate a gene, see what happens, and interpret the outcome.

From these experiments, it looked as though, in this organism, genes were directing the process of development, using programs embedded in the proteins the genes coded for. The interactions between cells and genes are so fast and the damage caused by mutations so profound that one can be forgiven for thinking that genes rule. Something in the back of my mind told me this was a mirage created by the fast pace of events. There were hints that cells were using genes for their own purposes and, in particular, that they were using signals—Wnt and Notch—to make choices about their own activities and to communicate with other cells during development.

It was not feasible to apply Feynman's advice and test our understanding by creating the organism we were studying. Flies have stem cells, but they are the adult variety, sort of like the ones in our gut; there are no fly embryonic stem cells—no pluripotent cells that can be grown in culture, differentiated at will, and coaxed into making eyes or brains. You cannot rise to Feynman's challenge with *Drosophila*. In this organism, in vivo means in vivo, and everything must be studied in the body of the embryo, the maggot, or the fly.

Then, around 2000, I heard about embryonic stem cells. As I learned more about what these cells could do, I began to think how they might be used to explore some of the most puzzling facts of biology. I considered Hans Driesch's experiments, described in Chapter 4, producing twin sea urchin embryos by teasing apart the first two cells—something that cannot be done in flies because of the unusual way in which nuclei divide first and cells only appear later, when there are thousands of nuclei. Most fundamentally, I wondered why embryonic stem cells could join pluripotent cells in an embryo to lay down the body plan for a mouse but couldn't be coaxed to do the same thing in a dish. Was some ingredient missing that, when present, would persuade the cultured cells that they hadn't actually left their original embryo, so that they would get back to work? I also felt that these embryonic stem cells would be a good experimental system to understand how cells use signals to build organisms. In the time that a fly embryo develops into a maggot, they divide once. With deliberation and a great deal of help and advice from Austin Smith, a pioneer in the stem cell field, in the early 2000s I moved my lab's focus away from flies and toward embryonic stem cells. At the time, there were just a handful of groups with a similar ambition in the United Kingdom, but I felt certain that these cells would help us find answers in our search to learn how cells make choices and transform these choices into the building of an organism.

We started with mouse embryonic stem cells. It was surprising to see that, when put in an environment that challenged the cells to make one of the layers that forms during gastrulation—be it the mesoderm (which eventually produces muscle, bone, and blood) or the endoderm (which eventually becomes the gut and lungs)—the cells would undertake the same steps as they would in an embryo. The genes that were expressed, as well as the schedules and programs of gene expression, were the same all the way down to the timing. This confirmed that hardwired circuits run programs down a mouse's Waddington landscape and that cells will follow them outside the embryo. It also served as confirmation that, as we have seen, in single cells, sequential interactions between transcription factors and the genome can create schedules, a form of time. Still, it was strange to see these sequences of transcription factors and fates emerge outside the context of a full organism and frustrating that the cells continued to refuse to assemble into an embryo or embryo-like structure.

The advances of other scientists encouraged us to keep trying. In particular, we were inspired by the news that Toshiro Sato and Hans Clevers had found a way to encourage embryonic stem cells to form an intestinal organoid, as well as by Yoshiki Sasai's observation of eye cups and brainlike structures built spontaneously by embryonic stem cells. One thing about these experiments caught our particular attention: the starting point in both had been not cells spread out on a dish but tightly bound aggregates of cells wrapped in Matrigel, that magic gluey substance that mimics the extracellular space. It was a vital reminder of the key way in which a dish fails to resemble life: there are no two-dimensional embryos.

Because embryos are built in three dimensions, we wondered whether something within the cells sensed the physics of mass and maintained an awareness of the numbers of its comrades before it would consent to generate a whole organism. Further, it appeared that the cells needed to be tucked into something that did a good job of mimicking the inter- and extracellular environments. That's why Matrigel seemed to play a role, but Matrigel itself was not as important as simply making the cells feel at home.

Of course, we were not the only ones asking these questions, and I came across an important clue to the puzzle in 2013, in a specialist journal called *Differentiation*.[2] While investigating how the initial organization of an embryo is established by activating *Brachyury* expression, Yusuke Marikawa and his colleagues at the University of Hawaii had observed cultured embryo carcinoma cells forming bean-shaped structures reminiscent of early frog embryos. In particular, in these structures *Brachyury* was expressed only at one end—the beginning of a sketch of a body plan. No Matrigel needed. This suggested that perhaps embryonic stem cells could be enticed to do the same.

I wrote to Marikawa to see if he had tried the experiment with embryonic stem cells. He had, but the cells had not organized themselves into the same kind of embryo-like structure. I asked two scientists in my lab, Susanne van den Brink and David Turner, to see whether we might have better luck. We did—but not in the way we were expecting.

MODELS OF THE WHOLE

One of the most wonderful experiences for a scientist is to conduct an experiment that turns out in an unexpected manner. I have often imagined the awe and amazement that Yoshiki Sasai must have felt the first time he watched an eye cup emerge from a culture of embryonic stem cells or the feeling that Toshiro Sato and Hans Clevers felt when they realized the intestinal stem cells they'd let grow for a few days had formed hollow bags of cells, a minimalist version of a gut. My team was aiming to make an "embryo." Instead we got something else—something that, over time, has perhaps proven even more interesting and eye-opening.

In the spring of 2013, using mouse embryonic stem cells and recipes we had developed to nudge embryonic stem cells to differentiate in culture, we managed to get them to form bean-shaped structures similar to the forms Marikawa had observed with embryo carcinoma cells. Like him, we could see the beacon defining the back end of an embryo, *Brachyury* expression, at one end of the structure. Not until we started filming what the cells were doing did I realize the crucial difference: they were trying to gastrulate.

To start the process, we had placed some embryonic stem cells in a small well, from which they could come together to form a spherical mass. Over the next two days, this mass of cells grew uneventfully, though sometimes a weak crescent of *Brachyury* expression would begin to show up in cells at one pole. On the third day, we sprinkled a chemical that activates Wnt signaling on the sphere; in response, the sphere became ovoid, with the cells expressing Brachyury leading the way. On one end, the cells grew and moved away until they had formed an extension of cells sticking out from the aggregate. From a point at the tip of this extension, cells continued to move; some kept extending outward while others circumnavigated the walls of the aggregate. Many of the cells within the aggregate looked the same, but by analyzing the genes each was expressing, we were able to map them to different types of cells, tissues, and organs. After six days, the aggregate had developed a tail at one end and heart cells, sometimes beating in synchrony, at the other.[3] There was a belly with the outline of a gut, a hint of a spinal cord opposite it, and the precursors of muscles and ribs. There was even a midline to the body, with the heartlike structure sitting to one side of the line. It wasn't perfect, but it was recognizable: a crude version of an early mouse embryo was forming before our eyes.

We called these structures *gastruloids* because they mimic the outcome of gastrulation: the organization of the body plan. Gastruloids tell us, unequivocally, that cells are the master builders, and there is no blueprint in the genome for what they do. If there were a blueprint, the embryonic stem cells would have made a mouse embryo as we set out to do—not some abstracted version of it, as a gastruloid is. Furthermore, gastruloids, perhaps more poignantly than organoids, show how an organism, with its proportioned and balanced organization, is more than the sum of its parts, more than the working of its genes. They are all pretty much the same size, and the domains of gene expression are in precise proportion with each other. In a flat culture, embryonic stem cells will become different kinds of cells depending on which signals are added, but they will never undergo the kind of organization seen in gastruloids.

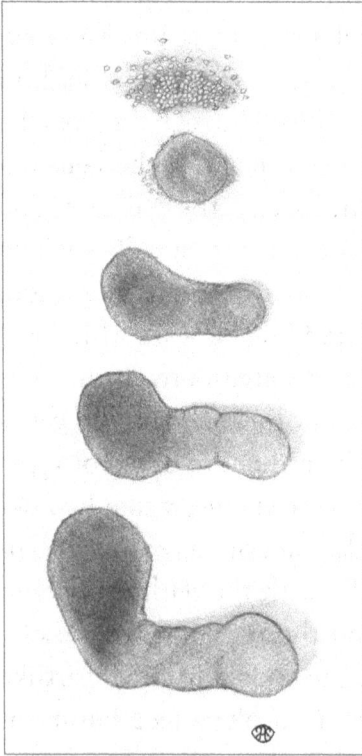

FIGURE 35. Emergence of a gastruloid from embryonic stem cells. The posterior end lies at the extreme of the elongating region (*bottom right*). Whether they are made from mouse or human cells, the outcome is similar.

Over the next several years, working with a small team of collaborators in Switzerland and the Netherlands, we began to explore how gastruloids come into being and to compare them with embryos. Despite being no more than 0.5 to 1.0 millimeter in length after five days in culture, the structures revealed many details about cell differentiation and specialization. Leonardo Beccari, working with Denis Duboule, the *Hox* master who first described the developmental hourglass of animals' universal body plan, showed that all of the genes of the *Hox* complexes are expressed in the gastruloids in the precise and correct temporal and spatial order seen in embryos during gastrulation. With Duboule's then colleague at the École Polytechnique Fédérale de Lausanne, Matthias Lutolf, we explored in detail how the cells of the gastruloid engineered themselves in three dimensions in the absence of external cues.[4] With Dutch biophysicist Alexander van Oudenaarden, we looked at the complex world of gene expression, coming to see how the cells deployed thousands of genes in precise patterns within their millimeter-long space and those patterns corresponded with those in the embryo.[5]

Gastruloids show that cells are able to organize themselves in space and create a coordinate system without instructions from the outside, as happens in the *Drosophila* or the mouse embryo. Furthermore, this ability to "self-organize" might be a fundamental property of cells, and similar, but less sophisticated, situations can be created in sponges and hydra: disaggregate their cells and let them aggregate at random, and they will re-form the organism.

Despite the level of organization in the cells' gene expression, at first sight the gastruloids lack that icon of organismal individuality: the primitive streak, though they generate a body plan. They also lack a brain, which is not entirely by chance. The signal we use to nudge the cells to start generating the gastruloid, Wnt, acts to suppress the formation of the seed cells for the brain early in development. Despite this, in the front portion of the gastruloid, we found scattered groups of cells with the gene expression signatures for parts of the head—a face without a brain. In essence, all organoids are disembodied sketches of our organs. In that context, it's not quite so shocking that cells can create a model of an embryo without a brain.

For years scientists had been aggregating embryonic stem cells into structures, called *embryoid bodies*, within which cells differentiated and specialized in a chaotic manner. What about the gastruloid led the cells to organize themselves in ways resembling an embryo? We discovered that there were two essential ingredients: the signaling regime described above *and* the initial number of cells.

It was particularly important not to use just any number of cells. Susanne van den Brink discovered that gastruloids will only form when a small but defined number of cells are used at the start. Following the Goldilocks principle, she began with too few cells, and nothing happened. She tried again with too many cells, and misshapen and confused structures emerged. Finally, she resumed with the "just right" number of cells, which in mice and humans happens to be somewhere around four hundred, and a gastruloid formed. When van den Brink increased this number by a bit, say by fifty, interesting things might happen, such as the formation of a conjoined twin. This number of around four hundred cells is intriguing because it's not very far from the number of cells at the time of gastrulation. Here was another sign that cells create space and time, using genes as tools and fixtures to help them accomplish their ambitions. The cells have the same genes from the start, but only when they amass enough numbers do they begin to reach into the genome to organize and build a body.

A distinct feature of gastruloids—one that still astonishes me—is that they show clear signs of going through somitogenesis, the process of forming somites, the primitive body segments that give rise to the trunk. Genes associated with the appearance of somites are expressed in the gastruloid in a precise order, from the back end to the front, just as in an embryo. Oddly, although the proportion of the total body mass of the gastruloid occupied by cells exhibiting the hallmarks of somitogenesis is the same as in the embryo, no actual somites form. Once again, we can see that gene expression isn't enough. The cells had the tools, but they've decided not to use them fully; they seemed to be missing something.

This is proof that genes don't instruct cells in how and when to build the embryo. Instead, cells respond to additional signals, mechanical and chemical words whispered from outside the cell, when deciding whether to pick up the tools and perform the job. We tested this by embedding gastruloids in Matrigel after three days. Just as they started to express somitogenesis genes, we watched in wonder as the cells came together to form somites, right where the genes had been expressed. Jesse Veenvliet in Berlin improved on this and elicited beautiful somites on either side of a spinal cord–like structure.

One of the other groups delving into the creative capacities of embryonic stem cells has been headed by a fellow professor at the University of Cambridge, Magdalena Zernicka-Goetz. A developmental biologist, Zernicka-Goetz has taken a different approach, exploring how a mouse embryo might be reconstructed from its basic components.[6] She started from the fact, well-known from genetic studies, that a mouse embryo gets its bearings through interactions with extraembryonic tissues: the trophectoderm, which contributes to the placenta, and the so-called extraembryonic endoderm, which gives rise to the yolk sac, an early feeding system for the embryo. Embryonic stem cells have a highly reduced capacity to generate cells of these tissues. This is why gastruloids lack the tissues that would support a connection with the mother—another reason it's surprising how much they can do.

Zernicka-Goetz's team placed the embryonic and extraembryonic cells together and added signals to get the process of development started. The result: some of the cells assembled into structures with a strong resemblance to mouse embryos about to gastrulate. In a small number of these cases, the cells managed to form a small primitive streak, establishing the beginnings of a front–back axis for the body, but they never developed a proper body plan. More recently, using variations in their experimental protocols, her group and that of Jacob Hanna at the Weizmann Institute of Science in Israel have managed to produce structures that, though imperfect,

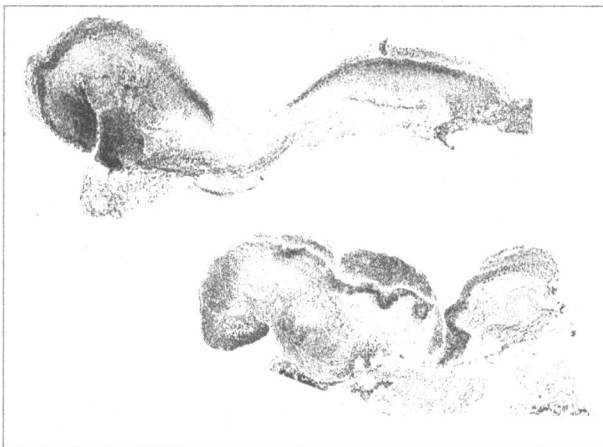

FIGURE 36. Natural embryos (*top*) compared to "synthetic" embryos (*bottom*). Despite the overall similarities, there are important differences in the details.

do look very much like embryos at the neck of Duboule's hourglass, exhibiting a full body plan with the primordia of a brain, a tail, and a primitive working heart.[7] They do not develop further, and for the moment the event only happens in a small number of the initial cultures, but the fact of it happening at all, without the intervention of sperm and eggs, is remarkable and underscores the ability of multicellular systems to self-organize.

That something very close to an embryo can be assembled in a lab from the three component tissues that would have made an embryo in the womb is remarkable and nearly rises to the level of accomplishing Feynman's scientific challenge. Yet, in my opinion, at the same time and for now, these assemblages do not tell us much beyond our existing knowledge of how embryos develop because they take us to where we started: the embryo. However, when they can be obtained in better numbers, I have no doubt they will teach us some new things as they will allow us to carry out experiments that are not possible with embryos.

Gastruloids, by contrast, remain full of surprises, like the fact that they find their pole despite not having their compass. How do gastruloids create a body plan without extraembryonic tissues to guide them? How do they sense the presence of the right number of cells, which allows them to organize themselves into the primordia of tissues and organs? How can we separate gene expression landscapes and morphology and how cells integrate both when presented with Matrigel? In part because of their imperfections, gastruloids have the potential to help us answer these questions. In the meantime, they offer proof that a confederacy of cells has the ability to work together, take cues from their environment and each other, and choose which genes to use and when.

BECOMING HUMAN: THEME AND VARIATIONS

Once we became adept at prompting mouse cells to make gastruloids, my colleagues and I naturally wondered if we could do the same with human embryonic stem cells. Despite the many similarities between humans and mice, our embryonic stem cells behave so differently that there was no guarantee we could just duplicate our mouse gastruloid protocol and succeed. However, Jianping Fu, an engineer studying stem cell mechanics at the University of Michigan, had managed to get cells to build structures with a similar arrangement to that of human cells right before the start of gastrulation, which gave us reason for optimism. Success in obtaining human gastruloids would be important. Suddenly, for the first time in history, we would get a glimpse of the processes that generate our body plan after the primitive streak appears around Day 14.

We immediately faced a major challenge in experimental design, however. This would be the first time we would have seen a semblance of human gastrulation in action. How would we calibrate the structures built by our stem cells?

Under a microscope embryos are, like their component cells, translucent. This contributes to their mystery, as it's quite difficult to see the internal structure of the specimens of early embryos that we do have, like those in the Carnegie Institution for Science's collection. Many of those embryos were sectioned and their sections used to reconstruct the internal structures, but today we have better ways to see and explore the interiors of those embryos. Recent technological advances now allow us to look at the organization of an embryo, cell by cell, minute by minute. At the Institut de la Vision in Paris, neurobiologist Alain Chédotal has been applying novel staining techniques to reveal the cellular structures of human fetuses in all their beautiful detail, going well beyond Lennart Nilsson's fanciful photo-essay. Chédotal's images have exposed how different organs emerge from their seed cells—the complex branching architecture of the developing lungs and blood vessels, the twists and turns that neurons take as they seek out communication partners and build functional networks, the majesty of the fetal brain and its eyes—but our identity is laid down much earlier, during gastrulation.

So we returned to the first principles of early human and mouse embryos. Remember, the initial events following fertilization are not that different. There's an initial proliferation of cells, which soon segregate themselves into the embryonic and extraembryonic tissues of the mouse. When the blastocyst is ready to implant itself in the uterus, the cells of the trophectoderm move around, burrow into the uterus, contribute to the placenta, and, together with the primitive endoderm, create a niche for the embryo. By the end of the second week, the cells of the embryo—about four hundred of them, as in the mouse—have organized into a disc, which is buried deep inside the extraembryonic membranes; this is different from the cup-shaped structure of the mouse at a similar stage. It is curious that the exception in terms of organization of the early structures is mice; rodents are the only ones to adopt the cup shape. The rest—pigs, horses, cows . . . us—organize a disc.

Then, as we saw in Chapter 6, on Day 14 or 15, a hint of the primitive streak appears at one end of the disc-shaped cell aggregate, and the dance starts.

As in the embryos of other species, human cells enact gastrulation by taking part in a multicellular choreography, lining up in a multitude of palisades that create the boundaries between different populations of cells, at turns folding and twisting to form tubes like the gut and chambers like the heart. By the end of the fifth week after fertilization, a human embryo has become an elongated mass of cells, about two and a half millimeters in length, with a clear body plan: a brain rising at one end, pushing the rudimentary heart down into position; a few somites, an incipient neural tube, and the outline of a liver and lungs in place; and a small tail at the end opposite the brain. By five weeks after fertilization, the rudimentary heart is beating sporadically, and blood cells can be seen spreading all around the embryo. Though it is amazing to picture such structures taking the form of a human baby in eight more months, the similarities to mouse embryos at the same stages are also striking, offering a vivid testament to our shared ancestry with other animals that goes beyond DNA.

Yet there are also essential differences. To start, in a mouse the process of gastrulation takes about a day and a half, while in a human it takes about six days. Other less obvious but important differences result from the fact that a mouse embryo begins by forming a cylinder of cells, compared with human embryos, which form a disc. We knew that many such differences would remain out of reach of observation and experimentation unless a human gastruloid could be obtained. My collaborator Tina Balayo started the process, but Naomi Moris, another member of my lab, made it happen in ways that surprised us once again.

It is not merely the shape of the embryo that differs between mouse and human but also the timing of events and the state of the embryonic stem cells. The tricks we'd developed to generate mouse gastruloids did not work with human cells, but by tinkering slightly with the signaling regime and adjusting the numbers of cells, we still managed to produce human gastruloids. It was a strange sensation to behold these multicellular structures, growing with so little intervention, and contemplate the knowledge that they were related to us.[8]

At first sight, human and mouse gastruloids are almost identical. Like mouse gastruloids, human gastruloids have a rudimentary heart at one end and a tail at the other, with different types of seed cells creating an abstract version of the embryo. They also lack the extraembryonic tissues that form the placenta, a primitive streak, and a brain—more cubism. When we analyzed which genes were being expressed by which cells, we observed that some cells, in the right place, were expressing genes associated with somites, and again there were no somites. We realized this provided us with a clue to pinpointing the moment in development that the gastruloids are equivalent to. We scoured the Carnegie collection for indications of the start of somitogenesis. Eighteen-day-old embryos don't have somites, while twenty-one-day-old embryos have a few, so our human gastruloids were equivalent to an embryo at Day 20 after fertilization.

Because many developmental disorders appear to arise during gastrulation, the creation of human gastruloids opens up exciting new avenues for research. We have shared the details of our protocol with researchers around the world, with variations made by labs to explore targeted questions.

A UNIVERSAL MAP OF CELLS

We had set out to produce embryos but got gastruloids instead. Obviously, these are not embryos; nor do they aim to be. I am biased but believe they are more interesting than embryos, because they tell us what pluripotent cells can do when they are trying to make an embryo without the help of the extraembryonic tissues. Surprisingly, they make the outline of an embryo, and they do it so robustly and repeatedly that they must be telling us some secret about their inner workings in the cellular engineering process that is making an embryo. Over time I have also realized something that I did not expect—namely, that gastruloids reveal something about evolution.

A very fundamental outcome of the process of gastrulation is the emergence of a system of coordinates that serves as a reference for positioning the precursor cells for tissues and organs and, with it, the body plan. As noted above, in mammalian embryos, this compass is created through a dialogue between embryonic and extraembryonic cells. The fact that in a gastruloid the embryonic cells can do this on their own, without help from extraembryonic tissues, still surprises me. It suggests that the ability of cells to self-organize into a bilateral body plan is a very ancient property, perhaps present in the cells of some of the very earliest animals, definitely in the Burgess shale. After all, we share a common ancestor with fish, frogs, and birds, and this line of descent implies we inherited not only vast swathes of our genes but also a range of cellular capabilities.

Inspired by the gastruloids that were growing in the lab, Vikas Trivedi, a researcher in my group, and Andrea Attardi, a master's student, working with my colleague Ben Steventon, took explants of cells from several very early zebrafish embryos, put them in culture, and observed what they would do. After a few hours, the cells had organized into a structure very similar to a mouse gastruloid—not only in the bean-shaped structure of the cells but also in the basic organization: heart at one end, tail at the other, with *Brachyury* expressed where it should be. When they presented their structures to me, unlabeled, I couldn't work out which was mouse and which was fish. It was as though Karl Ernst von Baer's embryo specimens had returned in the form of gastruloids. Reviewing the literature, I realized that, several years earlier, Jim Smith and Jeremy Green, then at the National Institute for Medical Research in Mill Hill, on the outskirts of London, had obtained similar structures from early frog embryo cells. Though these were never studied in detail, with hindsight we can recognize that they were a form of gastruloid. These observations suggested that cells have an intrinsic and impressive ability to organize themselves and that this ability is ancient and shared by many species.

Unlike the mouse embryo, where, due to the experimental difficulties, little was known about its early development, fish and frogs have a deep history, with much experimental work behind them, and there is a belief that molecular templates assist in the organization of the embryo. For this reason, the observations of Trivedi and Attardi needed to be tested to the extreme. This was done by Ben Steventon and his student Tim Fulton. They started by dissociating and aggregating the cells; the cells formed the same structure as the explant. They then shuffled the cells after dissociation and before aggregation; the cells formed the same structure again. They tested the organization of gene expression in detail and found a fish body plan.[9] On the whole, the experiments revealed, beyond doubt, that the cells of the early zebrafish embryo had a deep self-organizing ability resembling that of the mouse cells. The structures were called *pescoids*, and like the mouse gastruloids, they required a minimum and defined number of cells if the structure was going to emerge. Furthermore, the body plan revealed by the patterns of gene expression was similar to that of mouse gastruloids.

Pescoids were more surprising than the structures we obtained from human embryonic stem cells. Fish, frog, mouse, and human embryos look very different from each other at the very start and then again later, after gastrulation; remember Duboule's hourglass.

The differences at the start are the product of the distinct organization and properties of each animal's egg. The eggs exhibit different spatial geometries and exert different mechanical pressures on the cells whose descendants will form the embryo; for example, fish and frogs have large amounts of yolk, whereas mammals have extremely small amounts. Most probably, the width at the base of the hourglass, representing the beginning of development, reflects each animal's egg design and physical constraints.

In retrospect, perhaps we should not have been so bewildered with cells' ability to conjure up a body plan on demand. It happens in nature. Take a type of tropical fish called killifish that inhabits lakes that dry up each year. It survives where there is no water through the powers of its cells. During the early phases of killifish development, the embryonic cells disperse and enter a short state of suspended animation, called *diapause*, after which a few come together to form an aggregate with the body axes and, eventually, an embryo. If the cells happen to enter suspended animation right before the dry season comes and the lakes run out of water, the cells remain in their suspended, dispersed phase for as long as it takes for the rains to return and fill the lakes with water. Only then, through a mechanism or signal not yet identified,

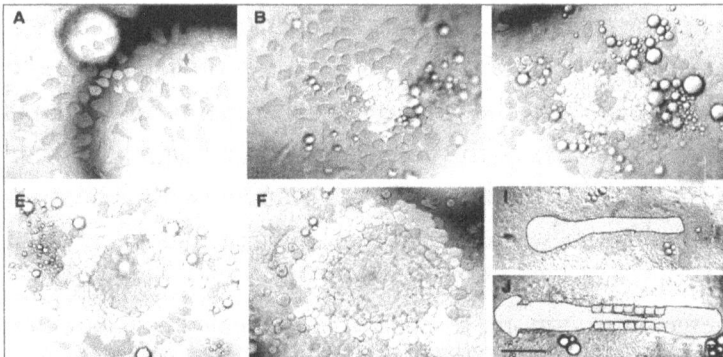

FIGURE 37. Early development of a killifish embryo. Cells are dispersed early in development but then come together in a small aggregate that undergoes a number of changes to form an elongated structure, resembling a gastruloid, from which the basic vertebrate body plan will emerge (*bottom right*).

will the cells come together and resume development. When you look at the aggregates in suspended animation more closely, they resemble a gastruloid.[10]

Taken together, these experiments suggest that by releasing embryonic stem cells from the constraints of the egg, as we do in our gastruloid and pescoid experiments, we get to see the cells assemble into a shape that is different from what nature would expect. Imagine taking a long balloon and twisting it into any number of animal shapes; you can accomplish this through force constraints, but when you remove those forces, the balloon will relax back into its natural form. So it is with cells in an embryo. Further, it appears that gastrulation is the period in development when the force constraints of the egg are removed. Somewhat unexpectedly, the cells, regardless of their animal of origin, produce the same shape, a simple, polarized structure with *Brachyury* expressed at one end, the rudimentary heart at the other, a midline, and two sides—in other words, the shape of a gastruloid.

These experiments revealed an underlying common pattern, a basic ground plan, clay for sculpting animals. Whatever genes do in the development of an organism, the evolution of life-forms is constructed by the geometry and mechanics of cellular ensembles.[11] Whether generated from fish, frog, mouse, or human cells, when freed from the constraints of the egg, cells build the same simple shape—a basic and conserved form that we have called the *morphogenetic ground plan*. Gastruloids are one incarnation of this plan, which I believe is common to all animals and probably originated in the days of the Burgess shale, when there was a huge explosion in animal diversity, with cells—not genes—trying out different constructions. As noted in Chapter 3, the defining event of life on earth, the emergence of animals from unicellular ancestors, rests on cells' discovery that they can work in ensembles to conquer and sculpt space.

The ability of early embryonic cells to self-organize stands in contrast to Lewis Wolpert's idea of positional information—at least his original one. This notion rests on positional signals being diffused across a field of cells that react to the local concentration; remember Sonic Hedgehog and the digits. However, when gastruloids organize themselves, there are no spatial references, and this raises questions about how cells find their position in the first place and whether cells can choose which signals to hear and which to ignore.

The seeds of an answer can be found in a very surprising piece of work by Alan Turing, of Enigma code-breaking fame, titled "The Chemical Principles of Morphogenesis," which he published the year before he committed suicide. Combining his genius and his knowledge of mathematics with intuition to understand how chemicals could be used to create the patterns that are the essence of nature, he began with a simple thought experiment. If you had two chemical substances in liquid form—say, one red and the other blue—and combined them, you would expect the substances to diffuse over time, producing a purple mix after a while. He then calculated conditions where the combination of chemicals could create stable spatial patterns—like leopard spots or zebra stripes. Importantly, for this to happen, the substances had to react with each other, interacting in defined ways, and only then would the patterns emerge. It sounded outlandish, but after Turing's death, his theory was shown to be true experimentally in several purely chemical systems. Later, German physicist Hans Meinhardt showed that within a field of cells, this mechanism could generate a coordinate system out of chaos. Ever since, variations on this theme have been used to explain how, in biological systems, patterns can arise from chaos.[12] Once a pattern emerges, positional information takes over and becomes the business of cells.

The bottom line is clear: we are all the result of cells interacting with each other in space. This places us—as Karl von Baer, Ernst Haeckel and Denis Duboule would want—firmly in the community of other organisms. It also challenges us to reconsider the special status that we give to ourselves and our embryos.

HUMAN EMBRYOS UNDER THE MICROSCOPE

Only about one-third of fertilized eggs, whether conceived in the womb or in a dish via in vitro fertilization, make it to become a fully grown baby. Among the lost pregnancies, most miscarry before or during gastrulation. Further, as many as six in every one hundred babies are born each year with a disease or syndrome, with less than half of these conditions being linked to any genes. Even those linked to the genome aren't *genetic* in the sense of their being associated with a specific mutation. Instead, they reflect how cells use the tools in the genome to build the body, and many problems in newborns are thought to result from errors in the execution of the choreography of gastrulation.

Very likely, a key cause of developmental dysfunction has to do with the loss of chromosomes during the formation of the blastocyst, a mistake in the gene-copying processes of mitosis during the early divisions that is surprisingly common in human embryos. Without certain tools and fixtures, cells simply cannot lay down and build the body plan. The first proof of this concept was set out more than a century ago by German zoologist Theodor Boveri, who, like his compatriot Driesch, was doing research with sea urchins. Boveri engineered zygotes such that, as they began to divide, they would lose chromosomes at random, resulting in sea urchin embryos whose cells contained different amounts of DNA. He then meticulously recorded the consequences. Cells' capabilities during development correlated with how many complete chromosomes they had; those with more complete chromosomes made a "better," more normal contribution to the embryo.

Although sea urchins have taught us a great deal about how cells use genes to build an organism, it goes without saying that these animals are very different from us. The same is true of the other species—fish, frogs, chicken, sheep, and even mice—enlisted over the past century in groundbreaking cloning and chimera experiments. However, we all share pretty much the same set of tools;

the differences lie in how cells use these tools and fixtures and what we build with them. Gastrulation again offers prime examples. For instance, in mice, having access to the *WNT3* gene is absolutely necessary for gastrulation to proceed. In contrast, for humans it's not, because humans lacking the gene still exhibit a well-laid-out body plan and normal coordinate system.[13] However, this doesn't mean that *WNT3* isn't absolutely required for normal human development; without it, fetuses fail to develop arms and legs. As another example, in mice, the transcription factor ISL1 is involved in the construction of the heart; without the right amount of it, mice develop cardiomyopathies. In humans, ISL1 appears to be associated with the construction of the amnion, the membrane sac that surrounds the embryo and fetus during gestation. There are no mutations in ISL1 associated with heart defects.

Small differences in specification matter, and in the building of embryos, which genes the cells use for specific tissues and organs can vary from animal to animal. For this reason, if we want to learn about human embryos and human developmental disorders and diseases, there is no substitute for studying human embryos, or something like them.

Of course, this is complicated by the special moral status that we bestow upon our own species. The Warnock Committee, when it established the 14-day rule, set out to protect unknowable individual embryos, most of which were never destined to survive development. Learning why some embryos do not come together in fully functional ways would be helpful, emotionally, physically, and financially. True, we can scrutinize the embryos in collections like the Carnegie Institution's for pathological specimens, but these provide a snapshot in time. There is much more to know. To dissect the exact nature of the relationship between cells and genes, to listen to the dialogue between cells that begins early on, when the blastocyst is forming, and to observe how and when the dance of gastrulation doesn't follow the choreography requires research that is currently forbidden. While gastruloids can answer some questions, they cannot stand in for embryos to answer all of them, particularly those associated with development before implantation.

There are solutions to this impasse. One lies in surplus embryos from IVF that can be used in research, after the donor's consent has been secured. Already, studies with these embryos have allowed scientists to observe the remarkable ability of a fertilized egg to develop into a blastocyst in culture. In addition, in many IVF clinics, every embryo that is implanted is filmed, which has helped scientists to understand the timing of cell divisions and the movement of cells during the creation of the blastocyst. They then look to see if features of this process align with whether the pregnancy succeeds or fails with the hope that in the future it will be possible to identify the viability of an embryo on the basis of its cells' behavior in the dish. If we understand the meaning of those movements, we can prevent a number of potentially damaged embryos from being implanted. In cases where there is a known genetic risk for the newborn, IVF clinics do preimplantation genetic diagnosis to select embryos that do not have one of a small set of specific disease-causing mutations, including those leading to cystic fibrosis and some breast cancers. Embryos are also checked for chromosomal alterations, such as those that cause Down and Patau syndromes. However, genetic and chromosomal problems account for a minority of embryos that have developmental challenges. One thing has become clear: many of the secrets to a successful pregnancy do not reside in our DNA.

In 2016 Zernicka-Goetz and another researcher, Ali Hemati Brivanlou at Rockefeller University in New York City, reported that they had managed to culture surplus IVF embryos in vitro up until gastrulation, at which point, they said, the experiment was stopped purely to stay within the bounds of the 14-day rule.[14] In their experiments, only a very small number of the embryos made it all the way to Day 14; however, those that made it were not in great shape. Nevertheless, the result was promising enough that many researchers lobbied for a change in the rule to allow experiments with surplus embryos in culture to continue, in hopes that the entire process of human gastrulation might finally be observed. Then, in 2021, the International Society for Stem Cell Research recommended that experiments with cultured IVF embryos be allowed to continue into gastrulation, but for how long remained unsettled.[15]

Some of the suggestions for an end point—for example, the first beat of the rudimentary heart or the emergence of brain cells or a working sensory nervous system—are likely to be contentious among nonscientists, just as they were among the members of the Warnock Committee. What does it mean for a heart to beat if it's not yet hooked up to a circulatory system? How can we know that nerve cells are capable of sensing pain if the embryo is maintained in a warm, comfortable environment? Regardless, scientists appear to be forming a consensus that research with surplus IVFs in culture should be allowed to proceed without defining a limit. We should test the cells to see how long they're willing to build the embryo outside a body, the thinking goes, in part because this might reveal aspects of human development that are shaped by the mother's womb. I agree with this in principle, even though at the moment there is no evidence that cultured human embryos will undergo gastrulation successfully in vitro.

Regardless of what embryos may or may not do in culture, two important issues are often glossed over in these discussions. One is the need to consider the number of IVF embryos being used in these experiments. That these embryos were surplus from IVF and would have been discarded if they weren't used in experiments is not, in my opinion, a defense for using them in large numbers. In all cases, the experimental design should be described in detail.

This leads to a second consideration, one that is potentially more significant for these studies. Recall that in normal circumstances, about two-thirds of embryos fail to complete gastrulation, ending the pregnancy in an early miscarriage. At the moment, none of the IVF embryos placed in culture for experiments are normal when they arrive at Day 14. How can we know whether the problem observed in the experiment was caused by the experimental variables or was the natural fate of the particular embryo? What are our *controls*, or points of comparison, for the results of IVF embryo experiments?

To answer these questions, we must better understand the process of implantation, which, in humans, clearly impinges on the pregnancy's viability, more so than does gastrulation. Studies of blastocysts in other mammals, such as pigs, which go through gastrulation without implanting, might help to tease apart the role of the embryo's location within the womb in delivering signals to cells that help them to organize themselves.

Growing an embryo in culture until Day 14 is not the same as growing an embryo ready for gastrulation, and growing a gastruloid is not the same as growing an embryo. Yet it's uncanny to watch the cells in these structures choose what to do and be. They are alive, but do they constitute an individual? We need to look into questions like this because behind them lie others that target what we are and the essence of our being.

– nine –

ON THE NATURE OF A HUMAN

THE RUBICON IS A SHALLOW STREAM IN NORTHERN ITALY, about ten miles west of the charming coastal town of Rimini. It is such a small and unassuming trickle that one might not realize how historically important it is. That's because one January day in 49 BCE, Julius Caesar broke a ban imposed upon him and his forces and crossed south over it, provoking war. It was a risky decision, with no going back, and Caesar knew it. According to history, the night before he crossed his literal Rubicon, Caesar muttered, "Alea iacta est" (The die is cast), to indicate the irreversible nature of the course he was taking. Luckily for Caesar, he prevailed, but ever since, the phrase *crossing the Rubicon* has been used to describe any significant, daring transgression.

The process of gastrulation is a Rubicon because, once it has started, the die is cast for the embryo's development; it can't go back (by natural means). We have spent much of this book talking about embryos of different animals, traversing this Rubicon that Lewis Wolpert deemed life's most important moment, and we have accepted this crossing without questioning its meaning and importance. An embryo is one of those things/objects/structures that, when asked to define it, we say, "I cannot tell you what it is but will point it out to you when I see it." We use the term loosely, assuming that we understand each other. However, when it comes to considering *human* embryos, something changes, and we cannot be hazy. Perhaps this is because of Anne McLaren's suggestion, inscribed in law in the United Kingdom after the Warnock Report, that at gastrulation the individual makes its entrance into the world. With us, the notion of an individual becomes entangled with that of an embryo, and we need to pin down an answer to a question that is easily taken for granted.

WHAT IS AN EMBRYO?

If you were to ask a random group of people—or even more precisely, individuals about to undergo in vitro fertilization (IVF)—What is an embryo? you would encounter a wide variety of answers: a baby in the womb, a fetus, an organism in the early stages of development, a blob, or, more likely than not, that plague of pollsters, the response "I don't know."[1] And those who hazard a definition will often append it with a "maybe" or "I'm not sure." This is not a failure of science education. It turns out that even among experts on development, the question does not have a straightforward answer. However, as we are about to embark on a journey to a new frontier in scientific knowledge, particularly with the emergence of embryolike structures from embryonic stem cells in the lab, we should start deciding where we stand and where we're going in this realm. There are, of course, political landmines in this labeling territory, but we need to press on.

Traditionally, meaning after the discovery of IVF, a human embryo was understood to be the product of the fertilization of the egg by the sperm. Cloning challenged this and has led some countries to develop more nuanced definitions that include the outcome of cloning. Others have left the definition as it was so that research could proceed with less hindrance than it would otherwise. The definition is important because, as we have seen, an embryo has rights.

In my view the term *embryo* refers to a multicellular structure that contains the outline of the body plan of an organism, whether a human or any other animal, with the precursors of the tissues and organs in place. It is the product of gastrulation, emerging through this process, as we have been discussing. Vitally, it should have *full organismal potential*—the capacity to generate each and all of the organs of an animal in working order.

One might respond to this definition by suggesting that the cells that give rise to the blastocyst, the rapidly growing ball of cells in early mammalian development that implants in the womb, should qualify as an embryo. Although the term is often applied to this structure, I don't think so. This mass of cells has the *potential* to become an embryo but lacks an outline of a body plan, meaning that it has not executed this potential—yet. In addition, as pointed out many times before, if split, it can give rise to two or three embryos; it has not individualized yet; individual means "indivisible."

That mass of cells needs to undergo gastrulation to execute its potential, and most of the time, the cells fail to do so before or during the process. This view agrees with that of the Warnock Committee in attaching great importance to the start of gastrulation and drawing a line at its onset to mark when a group of cells becomes an individual. Significantly, full organismal potential also requires the support of extraembryonic tissues—the placenta and the yolk sac— that allow the embryo to interact with the mother and go to term. For this reason, I believe that the presence of those supporting cells must be part of any meaningful definition of a mammalian embryo.

As a consequence of this definition, I believe, it is possible to separate the potential of a group of cells to form a full organism from the realization of that potential as these cells become an embryo, then a fetus, and finally a newborn.

These nuances are crucial to our understanding of human development because they help us talk about the identity and significance of the different types of cellular structures being created in labs from embryonic stem cells. For example, in summer 2021, two groups of researchers claimed to have obtained human *blastoids*, or blastocyst-like structures, from stem cells.[2] One of the groups, led by Jose Polo of Monash University in Melbourne, used human skin stem cells, inducing them to become pluripotent stem cells, to derive their blastoids. The university press release hailed it as "a game-changer in unlocking the molecular mystery of early human life." The researchers made clear that the blastoid wasn't a true embryo because it couldn't develop into a person, but nonetheless media around the globe announced that the lab had created an "embryo" out of skin, as though the researchers might quilt a human together. Like the stem cell–based structures in the last chapter, these bypassed the egg and the sperm.

No field of scientific research is free of such distortions of media hype. Just as Ernst Haeckel exaggerated the similarities between different embryos, today's pressures to obtain funding for research, augmented with a bit of human vanity, can lead some scientists to seek attention for their hard work from the media, which in exchange can—and often will—distort science, erasing subtleties and complexities, in pursuit of provocative headlines. When it comes to covering research that uses stem cells to make brain organoids or embryo-like structures, such hype can be particularly intense, because these experiments speak to our essence as beings. In general, but particularly on these topics, we need to be careful in communications with the press because if the reality does not match what is said, the surprise will erode social confidence in what we do.

Closer scrutiny of the Polo lab's blastoids, as well as those created by the other group, which used more conventional embryonic stem cells, revealed that these cellular structures differed in important details from their natural counterparts. In particular, the blastoids obtained from skin stem cells contained an unsettled mix of cells at different stages of development and didn't have any cells that could give rise to a placenta.[3] There was no way these structures could ever become a full organism.

Shortly after Polo's blastoid splash, two teams announced that they had succeeded in generating structures that were much more

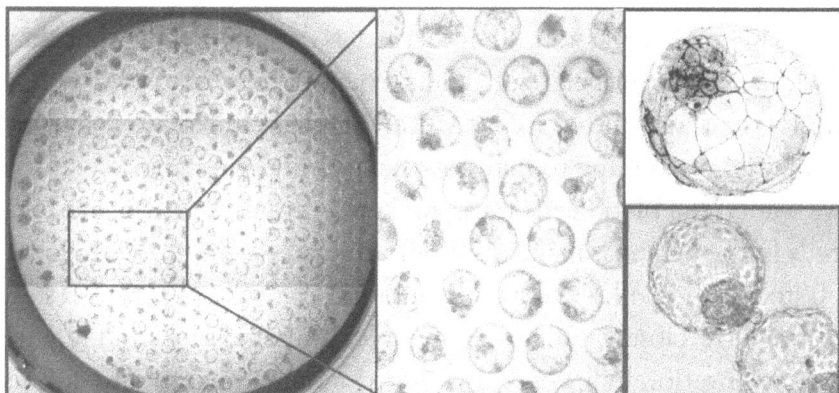

FIGURE 38. Human blastoids generated from embryonic stem cells. *Left:* arrays of human embryonic stem cells in the process of forming blastoids. *Right:* details of blastoids.

like the natural ones. The blastoids in the experiments performed by Nicolas Rivron, of the Institute of Molecular Biotechnology in Austria, and Austin Smith, now at the University of Exeter, contained all three main cell types that come together to create the embryo—and only those cells.[4] Curiously, the main difference between their experiments and earlier attempts was the starting culture conditions. On paper, this should not make a difference, as they were working with essentially the same cells, and definitely cells with the same DNA, but it seemed the cells changed what they did depending on the environment they were cultured in; this should not surprise you by now.

Whether Rivron and Smith's blastoids might have full organismal potential is another matter. To measure how similar a blastoid is to a blastocyst, researchers looked at their cell populations in terms of the genes they expressed. A blastocyst should have three and only three populations—the embryonic cells and the two extraembryonic cells—each of which expresses a distinct cohort of genes. Based on these criteria and on the organization of the cells in the blastoids, Rivron and Smith's structures remain as close as they can be to a blastocyst, and therefore, at face value, they maintain full organismal potential.

However, regardless of what cells, tissues, and organs look like in terms of gene expression, their identity is determined by what they do, their emergent behavior. Thus, groups of neurons conduct electricity to other neurons in a way that manifests in certain behaviors (walking, responding to light, or reaching and grabbing something we want), red blood cells transport oxygen to other cells, and beta cells in the pancreas make insulin. In the same vein, the cells in the blastocyst should lead to its implantation in the womb. As a result, testing how closely a blastoid resembles a blastocyst should require its implantation in a womb. But asking a human blastoid to do this as a test is not possible, for obvious reasons: humans are not experimental systems. Nevertheless such testing has been done with mice. Surprisingly, though mouse blastoids achieve a degree of implantation, they do not progress to gastrulation and instead are absorbed by the mother's body. Once more this demonstrates that an embryo amounts to something more than the genes its cells express; an embryo is what its cells do, and the uterus can sense this difference.

Going forward, should we want to test the activity and limits of human blastoids, we will need to develop a system that mimics or bypasses the womb and decide whether these multicellular ensembles have the same physical and moral status as a blastocyst. If blastoids become so similar to blastocysts that they are practically indistinguishable in what they do and become, should we treat blastoids in the same manner we treat blastocysts? Remember, the 14-day rule prohibits the culture of blastocysts beyond the initiation of gastrulation, but it is possible that in the near future, blastocysts might be able to do the transition to gastrulation ex utero. Then what should we do?

This highlights an interesting point raised by my colleague Naomi Moris, now at the Francis Crick Institute: Are blastocysts structures, or are they actually a process or product of a history that starts with the fertilization of the egg and continues until the emergence of the fetus—essentially embryos by the traditional definition? Is a blastocyst a mass of cells, regardless of how they come together, or does it have to be derived from a zygote in the womb?

If we err on the side of caution and decide that a blastocyst is a process, then blastoids are *not* blastocysts, because they do not arise from the same starting point. This allows us to afford blastoids a different moral status compared to their natural equivalents and might permit fewer constraints on their use in the lab. If, on the other hand, we conclude that a blastocyst is a structure, then blastoids and blastocysts should be ruled by the same principles. Under the current rule, that would mean that, barring new recommendations, once they begin to gastrulate, all research stops.

I believe blastocysts and blastoids are structures—structures whose history matters. This significance can be observed in experiments with mice, where, as we have seen, the womb accepts a blastocyst for implantation but not a blastoid with the same genetic makeup. It may be that blastoids do not implant properly because their cells haven't had the history of interactions with each other or their surrounding environment that a blastocyst has. In mammals, the mother's cells also help to make the embryo, and cells of the womb can tell the difference between a blastoid and blastocyst. The fact that blastoids do not arise from the fertilization of an egg is also relevant. Because their generation does not involve a zygote, it's possible to obtain large numbers of blastoids, derived from embryonic or induced pluripotent stem cells that share the same origin, and to compare them, providing a better experimental model for research. In theory, blastoids could provide an essentially unlimited source of structures for research use with the consent of the donors of the stem cells.

These are serious questions with much at stake and no clear answers yet. Blastoids could be used with great success in studies of fertility and early embryo development, but because they might become embryos, they demand sensitivity and the engagement of diverse viewpoints as we decide where to draw our lines. Furthermore, blastoids derived from induced pluripotent stem cells are as close to cloning as you can get. As with the Warnock Committee, I believe these deliberations should include scientists, doctors, religious leaders, politicians, lawyers, bioethicists, and the general public.

Commonly, the term *embryo* is used for any structure that arises after the fertilization of the egg, whether a blastocyst or a gastrula. I can see how this happens, and it may be a useful way of speaking without getting bogged down in awkward terminology, but when deciding what to protect by law, we need to be precise with definitions, and the recent advances in the use of embryonic stem cells to engineer the early stages of human development make this need more urgent. Though I am not a lawyer or an ethicist, as a scientist I have a view on these dilemmas formed through my years of research and observing others' experiments in this rapidly moving field. A blastocyst is a group of cells with the potential to form an embryo, but it is not yet an individual organism. For this mass to become an individual, it must first be transformed into an embryo. While the status of a blastocyst is very important, its transformation into an embryo, as defined above, is most significant.

This is because individuality is the notion that most challenges human sensitivities and emotions. The first article of the Universal Declaration of Human Rights, agreed on by the United Nations in 1948, states that we are all "endowed with reason and conscience," from which our rights as individuals flow. I agree with the Warnock Committee that before gastrulation, our cells do not have the rights of an individual, though this does not mean that they are not alive. Before gastrulation, our cells are alive, but they have not come together in the way necessary to make an embryo—as defined above—that would develop into an individual. For this reason, while we need to consider its potential when thinking about its rights and experimental possibilities, I believe that we cannot and should not treat a blastocyst as an individual. The same, therefore, applies to blastoids. Where I differ from the conclusions of the Warnock Committee is in when the individual emerges. The committee chose the onset of gastrulation, with the formation of the primitive streak. I see the embryo at this moment as still just a mass of cells; the primitive streak signals the initial organization of the body plan, but the body plan will not be complete until a few days later, well into the fourth week of development, when the embryo will arise.

Until then, cells are working together to establish their confederacy, and the individual is all potential.

STRUCTURES IN SEARCH OF MEANING

We haven't had the ability to observe human gastrulation in action for most of human history. The invention of IVF raised the possibility of seeing gastrulation in vitro, but almost immediately the 14-day rule reburied it deep inside the walls of the womb. Even so, tinkering with human embryonic stem cells in a manner that generates embryo-like structures has permitted us to cross a Rubicon in biology.

The announcement in August 2022 that scientists have generated a complete mouse embryo from embryonic stem cells—more precisely, creating the closest structure I know to a synthetic embryo—has drawn us into new territory.[5] However imperfect and infrequent their occurrence, these structures are very similar to embryos by the definition I have given above. Although they are made from mouse cells, in the next few years it will very likely be possible to do the same with human embryonic stem cells. This will be a momentous crossing of the Rubicon, and for this reason we need to be prepared, think about the ethical issues associated with these experiments, and deal with them before events overtake us. These structures will be helpful in understanding human gastrulation, but it has been suggested that they could also be grown as surrogates for organs and tissues for transplantation. This will require an extraordinary decision that, in my view, should only be taken if there are no alternatives, but I believe such alternatives exist.

Research on organoids derived from embryonic stem cells is proceeding apace and improving all the time, but it still lags behind what the organism provides—that is, the lab cannot imitate nature and produce organs with the structural and functional perfection that an embryo can. Organoids lack, whereas embryos provide, an environment of complex interactions that govern the relative positioning and assembly of emerging structures; this can be crucial for the development of certain organs—for instance, as we have seen, the heart. The group of James Wells at the Children's Hospital in Cincinnati has been working toward creating such arrangements to generate components of the gastrointestinal track; his group creates derivatives of the three germ layers involved in making up the intestine separately and implements the required interactions, bringing them together like a furniture kit.[6] The resulting structures maintain their autonomy but resemble more what develops from the embryo than classical organoids.

Another way to achieve those interactions is to let the cells do it themselves through self-organization. Here, gastruloids offer some possibilities by providing the means to observe gastrulation—or, more properly, the outcome of gastrulation—and thus can generate many interactions for free. Significantly, gastruloids are ethical because they do not have full organismal potential. They do not have a brain and lack the extraembryonic cell types required to attach to a mother. Without these supporting accoutrements, any embryo-like structure would find it difficult to develop very far on its own and certainly could not develop into a full-fledged organism. Even if gastruloids start to develop progenitor cells for brain tissues—and I am certain this will happen in the next few years—they will not be able to sustain their development into a fetus, much less a full organism, without difficulty. This is why gastruloids, on their own, are neither real nor synthetic embryos; instead they are *models of embryos*, or, to be more precise, stripped-down versions of the embryo. For this reason, I believe that gastruloids and related stem cell–based embryonic structures offer a solution to one of the great puzzles of modern biology: How can we learn what is and what isn't an embryo, or why the early stages of development so often fail while some succeed, if we can't run studies of embryos?

Gastruloids have thrown out some surprises and questions about the biology of embryos—about us. For example, we know that extraembryonic tissues forming the primitive streak can provide an internal compass that our cells use to lay down the body plan, top and bottom, left and right. How, then, do embryonic cells on their own line up in the same manner? If they can produce a body plan without a primitive streak, should this iconic feature of development be demoted? Anne McLaren and the Warnock Committee used the initiation of the primitive streak as the landmark for the emergence of the individual, but this reflected the knowledge and the prejudices of the time. Today, not only do we know more, but we have more experience with and information about embryos. Our views of the relationship between the primitive streak and the body plan have changed. For this very reason the International Society for Stem Cell Research has issued a welcome recommendation that the 14-day rule be relaxed.

Equally surprising is how gastruloids reveal the body plan. We know from extensive genetic experiments that embryonic cells use the *Hox* genes to change their form and function depending on where along the top–bottom body axis they are located. Gastruloids exhibit *Hox* expression that mirrors the spatial and temporal order in embryos, and therefore they will allow us to explore exactly how cells use genetic circuits to create time and space.

Because gastruloids follow the same cadences and patterns of development as embryos, they permit us to re-create and study pathologies with an early developmental origin in new ways. Importantly, by using gastruloids, we can reduce the use of animals in experiments testing *teratogens*, an important task that, at the moment, uses nonhuman models that are unlikely to reflect the behavior of human embryos. Together with other embryo models, they can also create models for disease. To do this, we will need to partner with patients, because it will be necessary to induce stem cells from a person with an identified disease before placing them in the protocol that persuades them to form a gastruloid. Already my colleagues Cantas Alev at Kyoto University and Naomi Moris are doing these types of experiments to understand the origin of familial or sporadic cases of congenital scoliosis and segmentation defects of the vertebrae, which are frequent in populations. Others will follow.

Studies like these are possible because gastruloids do not challenge current ethical guidelines. Nevertheless, we need to be cautious when making gastruloids from induced pluripotent stem cells, particularly those donated by patients. We do not want to replicate the heartache experienced by the Lacks family. Human gastruloids are neither clones nor replica pieces of the donor, but they do still retain a human, individual origin. We need to respect this by ensuring donors fully consent to the use of their cells to make embryo-like structures—nothing more but also nothing less.

As in geography, there are many ways to cross the Rubicon. Some of the paths we take may lead us to challenging terrain. I'm particularly intrigued by what embryo models will reveal about the creation of germ cells, those remarkable vessels that generate the next generation and carry portions of the genome along with them into the future. Germ cells are the *very first* specialized cells to be generated during gastrulation; this is probably an essential part of that Faustian pact between cells and genes. But in order for germ cells to fulfill their part, they have to be transformed into gametes—eggs and sperm—which involves interacting with gonadal tissue. Gastruloids make germ cells, but so far they have not developed gonadal tissue, something that happens late in development. Germ cells are a holy grail of biology not only because they are the path to the next generation but because the combination of sperm and egg triggers the magic of development.

Some daring experiments over the past decade have already begun to pave the way for discoveries that lie ahead. Among these are the work of the team of Mitinori Saitou at Kyoto University, which has managed to make functional mouse zygotes from mouse embryonic stem cells.[7] They have done this by deriving germ cells from embryonic stem cells and combining them with gonadal tissue to allow them to mature. Behind these spectacular experiments there is the hope that this research will lead to treatments for infertility. There are important differences between mouse and human germ cells, such as which proteins are involved in generating the cells, as pointed out by my colleague Azim Surani at the University of Cambridge's Gurdon Institute, who found some important differences between the two species—by now you will not be surprised by the fact that different organisms make use of the genomic catalogue as they can. For this reason, we are still far from generating human gametes from embryonic stem cells, but, to paraphrase Jacques Monod, what goes for the mouse, goes for the human, and research in this area is proceeding at an incredible pace. I wonder where it will be when this book is published.

Blastoids, gastruloids, and so-called *synthetic embryos*—often collectively referred to as *stembryos*—raise fundamental questions about who and what we are. How much of our development relies on the extraembryonic tissues that form the placenta and the yolk sac, which are consumed during or discarded at the end of gestation? Why is the success of germ cells linked to the success of the embryo? The results of the first cohort of experiments suggest that it will become possible in the future to steer embryos by changing their early environment—whether in vitro or in utero— to influence how the cells come together to form a unique life. What limits, if any, should we place on applying this knowledge? Once we gain greater experience with mixing and matching cells and with "fixing" the embryonic environment through the use of biochemical factors, critical questions will emerge that hinge on questions about who we are and what fundamentally defines us as a species.

ANCESTRY VERSUS IDENTITY

The new frontiers being explored with stembryos and organoids render the view that genes make us difficult to sustain. Genes are needed in development and tissue maintenance, but from the laying out of the body plan to the organization and functioning of our nervous system, cells rule their expression and make us who and what we are.

Throughout the twentieth century and continuing to the present day, the common assumption has been that our identity is tightly linked to our DNA. While there's some truth here—as Shakespeare said, "What's past is prologue"—when it comes to development, cells and genes have very different relationships with history. It is therefore worth pausing for a moment to recap briefly the reasons behind the dominant view of the genome as the master of our being. This view lies in the essence of DNA, those strings of Gs, Cs, As, and Ts that configure the hardware store catalogue that is your genome. This catalogue has been updated over millions of years and is unique to each of us, not in its repertoire of tools and materials but in their colors and design details. It is through these subtle differences that, in the same manner you can follow the transformation of the clumsy computers of the 1950s into your iPhone, you can trace history in your genome, a history based on the relatedness of the differences and similarities of the script. Add some historical narrative to these genetic relationships—genealogies—and you will have ancestries, lineages that, so you are told, link you to remote people and places and, if you go back far enough, link all people to each other.

We humans feel a great need to belong, to know our origins, and because we've been fixated on genes over the past hundred years, we've been using the language of genetics to write our stories. The big hitters in commercial DNA testing, Ancestry.com and 23andMe, together maintain the genomes of more than thirty million people in their books. Based on these data, you might be told you're 37 percent Western Bantu, 27 percent Germanic, 26 percent Scottish, and 10 percent Nigerian, give or take 10 to 20 percent, or that 2 percent of your DNA is Neanderthal (as it is, on average, for all people living today, because of how humans evolved). Some companies claim the ability to compare your DNA to that of, say, Vikings, ancient Egyptians, Chumash Indians, and other populations who lived a few thousand years ago. To be descended from a pharaoh! There's immense allure to imagining connections to such remote peoples, places, and times. But as the organization

Sense About Science says of such claims, "They are little more than genetic astrology." Geneticist Adam Rutherford has also rightly pointed out that if you go back far enough—and "far enough" isn't all that long ago in human history—we're all related to each other. This is a fact. The tools and fixtures in the genome are what we need to be an animal, a primate, a human, so it's not very remarkable that there's so much overlap.

Genetic ancestries can be quantified, saying that we are 50 percent this or 25 percent that, but what do these numbers really mean? Do these numbers say anything about who we are today? A history of our species may be carried in our genome, but our genome does not make us who and what we are.

An equally interesting view of human history is contained in the story of how our cells create us, division by division, starting from the zygote's very first division and building differences along the way. Recall the feature of embryonic development first noticed by Karl Ernst von Baer: that, just after gastrulation, we look very similar to a chicken, fish, and frog. The similarities in animal embryos connect us to the first multicellular organisms. Eukaryotic cells began to use the genome in novel ways, controlling genes so that they could come together to build a robust, efficient organism that could conquer the planet. To achieve this, cells found a handful of approaches that worked well, such as a bilaterian body plan and gastrulation. Across all animals, cells draw upon the same set of tools and fixtures to lay the foundation for the body. Then, after gastrulation, they go their separate ways, building species-specific features and unique, individual beings, based on their interactions with each other and, because we're mammals, our intercellular connections to our mother. The process of building does not end at birth; it continues throughout our lifetimes, so long as our stem cells generate new cells to keep our bodies running in good form.

Our genes aren't our identity, no matter what the DNA testing companies say in their advertising campaigns. Indeed, these

companies use the DNA from people's tests to research how diseases caused by a single genetic mutation might be treated. In other
words, they know that genes are tools, and they're looking for ways
to capitalize on the fact that these tools can sometimes be repaired.
And they aren't the only companies trying to profit from the misap-
prehension that we are our genes.

WHAT'S IN A TRAIT

In the summer of 2020, as we were all transfixed by the uncer-
tainties of living through the first global pandemic in a century,
Aurea Smigrodzki was born. She had been conceived with IVF,
but something distinguishes her from the long list of babies to
come into the world in this manner since Louise Joy Brown's
birth. There is something special about Aurea: she was selected
from among her parents' IVF embryos not simply because the first
cells were dividing well, or the blastocyst looked like other blastocysts that tended to succeed at implantation, but on the basis
of her genome. Of course, prenatal screening for genetic diseases
and chromosomal abnormalities is routine in medicine, with embryos checked for cystic fibrosis, Huntington's disease, sickle cell
anemia, and other conditions caused by a single gene mutation.
Aurea's screening went beyond this. She was selected because her
genome, it was believed, will give her good odds of steering clear
of heart disease, diabetes, and cancer as she grows up and ages.
She is the product of gene-centric thinking.

Aurea's selection was the work—one might want to say, *experiment*—of Genomic Prediction, a prenatal genetic diagnosis
company that has promised parents the ability to predict risk for
disorders including diabetes, coronary artery disease, and breast
cancer. The claim is based on the belief that statistical analysis of
DNA sequences can identify genetic variations across the entire
genome that, when found together, are associated with an increased
probability of the individual having a particular trait. Your proba-

bility of having the trait is called a *polygenic score* because the risk is based on many (poly) genes.

These risk scores are a relatively recent contrivance, a side effect of the publication of the first fully sequenced human genome in 2000. The publication triggered something of a gene rush to identify candidates for genes that cause disease. Immediately, genome-wide association studies (GWAS) were set up, with scientists searching across the genomes of many people to look for links between genes and conditions that appear to be inherited because they run in families. Beyond the obvious, already identified suspects, these studies never fulfilled their promise of simple associations of genes with phenotypes. Some conditions, including type 2 diabetes, coronary heart disease, and obesity, were believed to have a weak association with individual genes; however, further number crunching found that these correlations are most likely due to chance.

Ever ingenious, in the wake of these discouraging results, scientists posited that perhaps, more often than not, diseases do not result from one faulty gene but from a *combination* of defects in many genes, each making a small contribution to the disease. Nothing wrong with this since life is, as they say, complicated. Scouring the genome for these defects would be phenomenally difficult and time-consuming work because we have so many genes, and we don't know what most of them are or do. So instead, in this new variety of GWAS, scientists start with the phenotype—the exhibition of the symptoms of a disease or a trait—and look for any stretches of DNA shared by people with the phenotype. With the application of some statistical methods, the polygenic score was invented.

The polygenic score is not so simple as a calculation of risk based on observed patterns of DNA and diseases. A most important element of the calculation is the variation that exists in the structure of the genomes, the genotype, and what we see, the phenotype. And there's an assumption cooked into the score that it is the combination of mutations that is associated with the disease. This is because people in the population don't share 100 percent of all the DNA associated with the disease, and it's rare that 100 percent of the *heritability* of the disease—the amount of variation that can be ascribed to genes as heritable agents—can be linked to a suite of shared DNA. Thus, the score entails calculating what percentage of people share what percentage of the given regions and what percentage of disease incidence is genetic, then comparing this with how many of those sites you have in your genome to arrive at the odds that you might get the condition. It is technically complex and a mouthful.

Unsurprisingly, the interpretation of these scores is tricky. For example, say polygenic scores suggest that in white, Western populations, one hundred DNA markers explain 20 percent of the prevalence of a particular condition—that is, when combined, those one hundred DNA markers are associated with that condition 20 percent of the time. Turn this on its head: 80 percent of the heritability of that condition cannot be explained through the shared DNA that has been identified. Now, and here is the catch, these measurements say where you are in the broad spectrum of a population, but they cannot tell you what is going to happen with you as an individual. For this reason, acting on such polygenic risk scores, in my view, is problematic; these scores are not your fate.

There are other reasons to be critical, if not skeptical, of the usefulness of polygenic risk scores and other products spun out of genome-wide association studies. Both concern the data and their applications. For the most part, the data are derived from a largely white, affluent population living in North America and Europe. Practitioners who are fully supportive of using polygenic scores in screening and treatment always caution that the associations found so far might not extrapolate to populations of people beyond that on which the study was done. This isn't logically consistent. We share more than 99.5 percent of our DNA with everybody on the planet, though it is not always the same 99.5 percent. If where and how you grew up and live influence your risk score, that's proof the genome is a poor fortune-teller. Imagine if Isaac Newton had said, "Watch it! Gravity only applies to England because it is based on my observations of English apples." Of course, if we want to improve the predictive power of polygenic risk scores, we need only cast our net wider; if you put enough alleles into the calculus, you'll find an association between contracting diseases and being human.

In addition, the current risk scores are based on data from and applied to adults. These wide swathes of genes almost certainly participate in other processes in embryonic development, and selecting against them, or portions of them, may mean reducing the chances that the blastocyst will implant successfully or that some organ or tissue will develop fully.

Indeed, all of this would be a curiosity at the crossroads of statistics and biology were it not for a growing impetus to use polygenic risk scores in clinical settings. In the United Kingdom there is a plan afoot to sequence the DNA of every newborn and use polygenic risk scores to nudge them toward treatments before they exhibit any signs of disease. More companies like Genomic Prediction are sure to set up shop. Today they're gambling against health risk; in the future, intelligence, educational attainment, athletic ability, sociability, and longevity will almost certainly be in their sights.

Before we go down this particular path, we should consider the very real possibility of undesired effects. GWAS and polygenic scores are already being used in farming, to select breeds of specific value. For example, male broilers, or chickens selected and bred with one another to produce higher meat yield, exhibit higher levels of aggression than male breeders, which are selected to increase egg laying; foxes selected and bred with another because they have a mild temperament have offspring with a wider range of coat colors and patterns.[8] We just don't know all the ways in which cells use genes to make an organism. Thus, selecting human embryos on the basis of gene combinations associated with a disease, particularly one that is common in adults, might result in an unwanted trait, with little benefit.[9]

The DNA markers that define polygenic scores are not, for the most part, associated with particular genes; in many instances they are small footprints in the DNA called *single nucleotide polymorphisms* (SNPs). An SNP is a change in the DNA sequence with, often, no known association with any particular gene. This makes it all the more worrisome to read reports that polygenic scores are being considered for immensely complex aspects of our identity, like our personality and intelligence. We can speculate that these emergent qualities of our humanness are partly the product of thousands, or tens of thousands, of one-letter changes across the three billion letters of our genome; we know for sure that our personality and intelligence arise from interactions between our brain cells. There's no reason to favor the genome over cells here; for this reason, I would say that there is much to understand before we can trust polygenic risk scores to accurately describe any aspect of our psychological and emotional makeup, which are the product of interactions between our cells.

I find it especially dispiriting that when trying to locate the source of the risk for a trait that cannot be associated with anything genetic, people immediately look to the environment—the old nature-versus-nurture argument. *Nature* in these formulations means genome; *nurture* refers to the family and society. Over the last few years, the two sides have been conjoined through the field of epigenetics, which is merely a new kind of genetics. Cells have been "disappeared" from the debate. But you cannot get from genes to health (or to disease) without appreciating that cells are in between. A few—too few—GWAS studies sleepwalk in this direction by looking to make sense of SNPs by investigating the cellular behaviors that might be affected by small changes in DNA. The way forward, I believe, is not to translate statistical analysis of genes into a sense of a sealed fate but rather to appreciate the wondrous, dynamic activities of cells in continually and continuously making and remaking us. Maybe in this manner we can give meaning to these numbers.

THE HARBINGER OF FATE

Reading our genome is no easier than reading tea leaves to tell the future. This is because much of what we're looking for simply is not there. Your future is in your cells; much of it is actually in your stem cells, that special class of cells that ensure you get a new lining for your gut every week and a new skin every month. What is more, as we have seen, the genomes in each of your newly generated cells are almost certainly different from the ones that were in the cells that came before. With every new cell, there is a chance—a low chance, but a chance—that it will have picked up a mutation that causes the genome to break its pact with cells and try to take over. Most of the time, the cell will take care of the problem, by committing suicide or sending signals to the immune cells, which can help control the situation.

We have grown accustomed to the idea that mutations in our DNA cause cancer, but sometimes the harbinger of doom comes in another guise. Consider the case of Olivia and Isabella, identical twin sisters from Bromley, England. Both sisters were born healthy and happy. Then, not long after their second birthday, Olivia became ill with acute lymphoblastic leukemia, a cancer of the blood and bone marrow. This leukemia is known to be linked to a mutation, a fusion of two genes, *TEL* and *AML*, which both twins carried, but which on its own is not sufficient to cause cancer. This made the twins' parents and their doctors worried that it was only a matter of time before Isabella also showed signs of the disease—after all, as identical twins, they had the same DNA— but she did not.

In order to understand Olivia's cancer and monitor Isabella's health, their doctors conducted further DNA analysis. Oddly, when the parents' DNA was tested, neither mother nor father had the *TEL-AML* mutation, and the mutation was only present in the twins' blood cells; it wasn't in any of their other cell types. Moreover, Olivia's blood cells had an additional mutation—one she didn't share with Isabella—and that mutation tipped her blood cells into leukemia.

With some detective work, Tariq Enver, then at the University of Oxford, and Mel Greaves of the Institute of Cancer Research in London, found the explanation.[10] The girls had indeed been conceived healthy; the zygote that would become Olivia and Isabella had not carried either cancer-causing gene mutation. Then, at some point shortly after gastrulation, while their body plan was being constructed, the *TEL* and *AML* genes fused in a blood stem cell in Isabella. Because the twins had shared circulation, that stem cell passed on to Olivia and established itself in her bone marrow, where it underwent the second fatal mutation. Ultimately, Olivia's leukemia was caused by the two mutations, but they lay not in her genome but in the malfunction of one of her stem cells, which had its origin in her twin sister's body.

Scientists are learning more about which cells are able to stop the genome from breaking the Faustian pact. One of the most promising therapies being tested today, called *chimeric antigen receptor T cells* (CAR-T), makes use of the fact that many cancer cells have specific proteins on their surface; these proteins are like camouflage, allowing the cells to move around the body and cause havoc while evading immune cells like T cells. CAR-T cells repurpose a T cell to pick up a molecule that recognizes the proteins on the surface of tumor cells. To make CAR-T, doctors extract T cells from a patient's blood, tinker with them so they recognize the protein on cancer cells, grow them in culture, inject them back into the patient, and let them do their job. It works. Many patients have been cured with CAR-T.

In these two examples, cells are the harbingers both of ill and of hope. In the case of Olivia, her leukemia can indeed be mapped to mutations in two genes that, we could say, are the cause of the disease. We can also look at the CAR-T cells as the outcome of engineering the genome of T cells. In both cases, we could look at genes as the center and the cause of disease or cure. However, this is the same as if we looked at a craftsperson tinkering with his or her toolbox, updating and sharpening some tools and blaming tools for what they produce. Ultimately it is the user that makes sense of the tools and uses them to build objects creatively. In our case, the user is the cell.

CODA

At the end of the nineteenth century, biologists had recognized the cell as a fundamental building block of organisms. However, with few exceptions, they did not see it as an agent of space and time. Instead they saw the cell as a passive object, a brick that obeyed the designs of the tissues and organs it was part of and, ultimately, of the organism, which somehow "knew" what it needed to build itself.

This was certainly the interpretation that Hans Driesch and Hans Spemann derived from their experiments: the organism manipulated and ordered the cells; the organism ruled. Driesch likened the organism to a "machine" and, even when he found himself dumbfounded by his observations of the cells of the early sea urchin embryo, responded by trying to reason out what kind of machine might be able to perform the feats he had observed. Remember his experiments in which early on, each cell, when on its own, would give rise to an entire organism rather than just the part it would have contributed to the whole. He concluded a machine with such behavior could not exist. It would be, he mused, "a very strange sort of machine which is the same in each of its parts." He could not find an answer to his dilemma, a mechanistic answer, that is, and in despair turned to vitalism.[11]

Driesch lacked our current vision of the cell. In those early days, few could foresee that, behind those inert bricks revealed by the microscope, there was such a universe of activity and agency. Those parts of the machine that are the same and able to reconstruct the whole are the cells, with their hardware store imprinted in their nuclei. The other piece of the puzzle he lacked was emergence, that property whereby the components of the cell interact with each other and their environment to create a whole bigger than the sum of its parts, resulting in the many behaviors, actions, and spatial and temporal organizations that underlie the structure and function of organisms.

In the vision I have outlined here, the genome contains the code for the parts, tools, and materials that, when brought together within the confines of the cell and used in a selective manner, acquire properties that they don't have on their own. The cell is then able to use their interactions to control the activity of the genome for building and function. The genome may, just may, be responsible for building the first cell, the zygote, but once that cell divides to become two, a new world opens up in which the interactions of those cells control the activity of the genome in the conquest of space and the control of time. The discovery and acknowledgment of these emergent properties has transformed our vision of the cell from a static to a dynamic entity.

We are just beginning to understand how this works, and we are both reinventing the cell and creating a new science of the cell. Cells are no longer perceived as the static structures with different names that we learned about in school and cytology classes in college; they are complex, dynamical, and creative entities with the ability to learn, move, and count, to measure space and time, while communicating with each other. Nowhere do we see this in action more clearly than in the process of gastrulation and in the way cells build tissues and organs in culture, agnostic about the details of their genetic programs. In order to communicate with them and steer their powers in specific directions, we use their language, the same chemical and mechanical signals they themselves use to build organisms. It is true that we also catalogue the genes the cells express, but in the same way that we think of our DNA as a barcode for our identity, the expression of these genes is a barcode for the cells: each cell type expresses a repertoire of genes that can be used to identify it. But we should remember that cells are so much more than the genes they express; they are what they do.

At the dawn of the twenty-first century, three prominent biologists articulated an argument for the future controversially titled "Molecular Vitalism."[12] In it they hinted at a dynamic vision of the cell, from its molecular components to its behavior in developing an organism. In a veiled manner their essay is a call to arms to embrace the study of emergent properties of biological systems in the certainty that they will provide an understanding of how biology integrates the different scales into the wonder that we call the elements of nature.

Wherever you look, whether at yourself in a mirror or the vista of a forest, what you see is the creation of cells. The beating of your heart, your thoughts and emotions, and your ability to read these lines are linked to neurons, their electrical activity, their conversations and cooperation. The livelihood and survival of your gut, blood, and skin, your ability to run, to write and grab, and to continue to do all this for a long time, all these also depend on cells, but this time of a special brand: the stem cells whose activity will, in more ways than one, determine your health. And of course, your future in the form of the next generation depends on those special vessels, set aside early in development to pass on the gene to the next generation that we call gametes. We are part of an ongoing story that begins with the appearance of the first eukaryotic cells billions of years ago and continues with the discovery of multicellularity and the exploration of space and time forged in fungi, plants, and animals. From the moment our zygote was formed, we became part of that story, and our future, like our past, belongs not to our genes but to our cells.

EPILOGUE

At the turn of the twenty-first century, we take one last wistful look at vitalism, only to underscore our need ultimately to move beyond the genomic analysis of protein and RNA components of the cell (which soon will become a thing of the past) and to turn to the "vitalistic" properties of molecular, cellular and organismal function. . . . The genotype, however deeply we analyze it, cannot be predictive of the actual phenotype, but can only provide knowledge of the universe of possible phenotypes.

—M. Kirschner, J. Gerhart, and T. Mitchison,
"Molecular Vitalism"

I think we are unearthing some very important concepts. I am not sure that we will get to a set of macroscopic laws with which to understand biology. However, PF Lenne was right about us getting nowhere with focusing on the microscopic. Since the advent of molecular biology and genetics in the study of development, we have treated experimental embryological work as a bingo sheet with which to tick off the role of different genes/molecular pathways—hence

the focus on the microscopic. We really need to go
back to these concepts and attempt a new synthesis
and I think this is what we are starting to do. Modern
techniques hold the potential to make a huge leap
forward in the understanding of development.

—BEN STEVENTON, CORRESPONDENCE, 2018

I N 1864, WHILE CHARLES DARWIN'S STILL RELATIVELY NEW
ideas about evolution were being hotly debated, British prime
minister Benjamin Disraeli decided to weigh in on the discussion.
The idea that we might descend from apes shocked and revolted
many people, particularly in the context of a society with a solid
religious foundation. "What is the question now placed before the
society with a glib assurance the most astounding?" he asked. "The
question is this—is a man an ape or an angel?" He was met with an
uproar of laughter from his audience. "My lord," continued Disraeli,
determined to deliver his punchline, "I am on the side of the angel."
Today, at a time when so much of the debate is framed in binary
terms—Are we made by our genes or by epigenetics?—I must follow
Disraeli's formula: I am on the side of the cells.

The twentieth century was the century of the gene. It dawned
with the rediscovery of the work of Gregor Mendel and confirmation that the essence of heredity lies in discrete units of biological information passed on from one generation to another. As
the century progressed, an exhilarating sequence of discoveries
placed those units into chromosomes, showed that they could be
altered, or mutated, and that some of these changes are linked to
our health. Most significantly, genes were shown to be made up of
DNA within that iconic double helix. This was followed in quick
succession by the elucidation of the genetic code and of the mechanism that translated genes into proteins and how these perform
functions from carrying oxygen around the body to configuring a
cytoskeleton. Toward the end, genes were linked with development; the century closed with the unveiling of a draft of the hu-

man genome and a sense of triumph that we now could read the "book of life"—and, more recently, even rewrite it. These discoveries invited exalted claims that we now held "the complete set of instructions for our development, determining the timing and details of the formation of the heart, the central nervous system, the immune system, and every other organ and tissue required for life," as Charles DeLisi once said.[1] With such an amazing story to tell, it is little wonder that the gene has exerted such a spell on us. But as we have seen, the genome is not actually a blueprint for an organism or its architect. Insofar as it contains any design, it is the design for another genome, not for an organism.

It would, of course, be foolish to argue that genes have nothing to do with who and what we are; they do. But they are not the masters of our being and fate that they have been made out to be. The notion of a toolbox is often bandied about without ever answering the question of who or what is selecting and using the tools. As we have seen here, that elusive entity is the cell.

Notwithstanding these questions, the gene-centric view has become deeply ingrained, establishing a form of tyranny where genes reign supreme over not only our past and our present but also our future. At one extreme of this mind-set, psychologist and geneticist Robert Plomin has said that pretty much everything about who and what we are, and who and what we will become, is written in our genes from the moment of our conception.[2] He has suggested that social interactions or environment can do little to override the power of genes; we can only acknowledge our genetic selves and work around them. Such views are a natural extension of the idea that the genome contains our operating instructions.

Yet, as we have seen, without a cell, a genome doesn't mean much. For creatures ranging from a virus to a human being, it is cells that give meaning to those sequences of nucleic acids by translating stretches of them into proteins. It is cells that use those proteins to take care of and repair themselves. Most importantly, it is cells that work with other cells to construct an organism. The cell decides which genes are used for what purposes and when rather than being at the mercy of the genes, a feat on most magnificent display during the development of an embryo.

Late in the nineteenth century, it was established science that the cell was the fundamental basic unit of biological systems. The consequences of this realization were ignored, however, first because of a lack of understanding of how cells worked and later because of our obsession with genes. This is finally being righted with the discovery that we can coax cells to build embryo-like structures in a lab, without tinkering with their genomes, just using their language: BMP, Wnt, FGF, Shh . . . to communicate with them and steer their actions where we want.

It's remarkable, really: if you grow cells on a flat surface, they will spread out or round up, depending on the culture in which they are being grown, maybe even following programs of gene expression and adopting different cellular fates; but they will not engage in making an organ, much less an embryo. Place the same cells in three dimensions, and depending on the initial numbers, they will generate either chaos or an embryo-like structure, weaving sheets that they can mold into different shapes—the tubes of the gut and spinal cord, the chambers of the heart, the furls of the brain. When we obtain embryo-like structures, we are able to see why cells with the same genes use those genes differently, creating different spaces in different time and thus building the various tissues and organs of which we are made. Same genes, different outcomes, depending on the cells' immediate environment. Out of the interactions and communications of trillions of cells, we emerge. The cell is the architect, the master builder.

A critic might protest that by highlighting the power of cells over that of genes, I am endowing cells with mystical abilities that do not help advance our understanding of Life any more than reductionist genetics. And it's true that these are early days in our understanding of the workings of groups of cells, how they contribute to gastrulation, to the building of an arm or a heart. But it is clear that we are not going to make progress simply by cataloguing the genes they express; we need to engage with the emergent properties that give rise to cells and that arise from the workings of cells, find the elements that drive them, and learn to control them. Cells can't always be easily enumerated, measured, and compared, in the way that DNA and gene mutations can, but we do have some techniques for observing cells' activities, particularly how they communicate and coordinate with each other. The electrical activity of networks of neurons can be recorded in electroencephalograms and other scans; the performance of the heart can be monitored through electrocardiograms; the work of the immune system can be measured in specific outputs as body reactions. While we currently lack comparable techniques for monitoring the activities of cells in embryos and our tissues, let alone for quantifying how cells generate space and time during embryonic development, we are learning.

As we study embryo-like structures and come to better understand the operations of our cells, we will be able to explore in more detail the nature of the relationship between cells and genes and write new pages in the history of biology. It may be that cells lose, or cede, control to genes in cases other than cancer. What a remarkable thing that would be: a Faustian pact that is continuously being renegotiated in each living organism. But until we recognize the power of cells, such dynamic aspects of biological systems will remain invisible to us.

Based on everything we have seen in the behavior of stembryos and organoids, I feel certain that cells hold a creative potential that genes cannot dream of. Whereas genes provide a substrate for transcription and replication, cells display a broader repertoire of activities in the versatile and complex work of proteins, when sculpting tissues and organs into embryos and fully fledged organisms. It is often asked how such similar genomes can build such different animals as flies, frogs, horses, and humans. However, the real wonder is how the same genome can build such different structures as eyes and lungs in the same organism. Let us give cells their due.

Taking a cell's-eye view of life may at times feel messy. It will definitely be messier than the digital, abstract view of ourselves that we get from studying our genes, but we should remember that this is the start of a new page in the history of biology and that, as in other pages of science, there will be some fog in the beginning. Cells are intuitive and social, sensing and reactive to their environment in a complex, emergent manner. Their actions are not a simple matter of turning off or on switches.

A cell's-eye view of biology will provide a rich understanding of our being and our past. It will articulate the tussle that went on when animals appeared on the face of the earth, the tension between selfish genes and the intrinsic cooperative nature of cells. This was resolved with the cells taking control of the genome to explore the creativity inherent in their powers and creating a divide—the germ cells—for a safe passage of the genes to the next generation, where the story repeats itself. Looking at us from the perspective of cells brings us closer to other animals—far closer, at the neck of Denis Duboule's hourglass, than the overlap in genomes—and the uncannily similar sketch of the early body suggests a grand design to life that we are just beginning to uncover.

One senses a shift in our understanding of how we are made and who we are to a position in which genes, rather than determining every detail of biology, are integrated into the activity of cells. I can see a future in which a cell-based understanding of biological systems promises to help us tackle diseases and improve our lives with even more benefits than are being afforded by our current understanding of the gene. We can get a glimpse of this in the success of immunotherapy, where immune system cells are trained to hunt down and destroy tumors, as well as in the promises ahead in understanding how cells age and how this process can be reversed. As cells spill their secrets, revealing the ways in which structure and function develop side by side, the possibilities for regenerative medicine will be nearly boundless. We do not yet know much about how cells come together to use the genome, but the answers are out there, starting to come out in the workings of our embryo-like and organoid cellular marvels. The century that is now well underway is, and will be, the century of the cell.

ACKNOWLEDGMENTS

I HAVE BEEN INVOLVED IN RESEARCH FOR OVER FORTY YEARS, BUT research tends to give a biased view of the subject under study, restricted by the rules of science: objective data collection, analysis, and interpretation. This is as it should be, but there is more to it, and this often comes out when we talk to students, colleagues, and friends about our results and experiences. Then we are allowed, sometimes compelled, to search for the meaning of our work in a broader context. Whether by conveying our enthusiasm for the subject matter or in speculating about our findings, those discussions can uncover surprising meanings and implications of our work. This book reflects this and was brewed through many years of teaching and tearoom musings in this spirit.

During forty years in the United Kingdom, first in the Department of Zoology and later in the Department of Genetics at the University of Cambridge, much of my mind has been absorbed by how organisms build themselves from single cells. I have been lucky both that this time has coincided with a revolution on this subject and that I have been a privileged witness to these developments. This has led me to face the central role of the cell in the building of organisms and the reality that, despite this obvious fact, current explanations of the process are based on genes. This book is the result of this realization. My initial aim was to convey the beauty implicit in the development of animal embryos and the history of how we have learned about this amazing process. However, as I developed this theme, it became obvious that embryonic development is about a deep relationship among genes, cells, and organisms that, when explored, challenges the current gene-centric narrative of biology and our identity. Discussions with my editor, Robin Dennis, reshaped what the book was about, and I am very grateful to her for this and for teaching me to write in a way that, I hope, is broadly accessible.

The views presented here are my own, but they have been molded by many people, books, and conversations over the years. Interactions with Michael Bate, Denis Duboule, Jeremy Gunawardena, and Ben Steventon have been particularly important as they have taught me what I do not know and how to find a way toward an answer. The arguments on genes, cells, and organisms (and more) owe much to many hours of discussions—for the last three years on Zoom—with Adrian Friday, who over more than twenty-five years has patiently instructed me on the "gene's-eye view of life" by challenging my reluctance to accept it. Adrian has read and commented on every page of the manuscript, though this doesn't mean that he is responsible for its contents. I am very grateful to him for sharing his enormous knowledge about biology and his time.

Thank you to the many students and members of my research group who have contributed to this book unknowingly, by asking questions or sparking and encouraging thoughts that drove my mind to interesting places. Thanks also to Michael Akam, Ramiro Alberio, Paola Arlotta, Buzz Baum, Jaume Bertrantpetit, James Briscoe, Antonio Garcia Bellido, Jordi Garcia Ojalvo, Nicole Gorfinkiel, Jerome Gros, Kat Hadjantonakis, Nick Hopwood, Pierre Francois Lenne, Matthias Lutolf, Juan Modolell, Naomi Moris, Arcadi Navarro, Martin Pera, Andreas Prokop, Nicolas Rivron, Iñaki Ruiz-Trillo, Steve Russell, Aylwyn Scally, Christian Schroeter, Marisa Segal, Austin Smith, Shahragim Tajbakhsh, David Turner, and John

Welch for discussions and comments on various sections of the text. Bernadettte de Bakker, Miguel Concha, Madeline Lancaster, Prisca Liberali, Matthias Lutolf, Jenny Nichols, Giorgia Quadratto, and Nicolas Rivron provided images to illustrate some of the points of my arguments.

I also want to acknowledge the influence of Lewis Wolpert through many inspiring conversations and lectures. He was a leading light in developmental biology, an inspiration and an example of how to communicate science to nonscientists and, often, also to scientists.

This book would not have been possible without my research, the source of questions and the opportunity to search for answers. For this reason, I am especially grateful to the University of Cambridge, the European Research Council, and, more recently, ICREA for supporting my interests in a manner that has allowed me to develop projects that are at the heart of the ideas in this book.

I am particularly grateful to my agent, Jaime Marshall. This project has taken many zigzags, but he thought it worth pursuing and gave me the opportunity and support to make it happen; thank you to John Inglis, friend of many years, for introducing me to Jaime and thus setting the process in motion, and to Thomas Kelleher from Basic Books for believing in the project and encouraging me to explore territories where I was shy to tread. Brandon Proia has done a great job of editing the final manuscript, and Selma A. Serra has transformed my often abstract visions of sections of the book into enticing illustrations.

Finally, thank you to Susan Gatell, without whose support this book would have never been written and whose vision and excellent editing and writing skills played very important roles at critical moments of the project.

Writing a paper or a book is often a question of mood, environment, and inspiration, and this project is no different. It started in Morges, on the shores of Lake Lehman in Switzerland, during a sabbatical with Matthias Lutolf; it was continued in Cambridge, United Kingdom, and finished in Barcelona, Spain. Each of these places has left its mark on the stories and thoughts herein.

NOTES

CHAPTER 1: NOT IN THE GENES

1. J. Li et al., "Limb Development Genes Underlie Variation in Human Fingerprint Patterns," *Cell* 185 (2022): 95–112.

2. E. B. Lewis, "A Gene Complex Controlling Segmentation in *Drosophila*," *Nature* 276 (1978): 565–570.

3. E. Wieschaus and C. Nüsslein-Volhard, "The Heidelberg Screen for Pattern Mutants in Drosophila: A Personal Account," *Annual Review of Cell and Developmental Biology* 32 (2016): 1–4.

4. W. J. Gehring, "The Master Control Gene for Morphogenesis and Evolution of the Eye," *Genes to Cells* 1 (1999): 11–15; P. Callaerts, G. Halder, and W. J. Gehring, "Pax-6 in Development and Evolution," *Annual Review of Neuroscience* 20 (1997): 483–532.

5. T. J. C. Polderman et al., "Meta-analysis of the Heritability of Human Traits Based on Fifty Years of Twin Studies," *Nature Genetics* 47 (2015): 702–709.

6. R. Joshi et al., "Look Alike Humans Identified by Facial Recognition Algorithms Show Genetic Similarities," *Cell Reports* 40 (2022): 111257.

7. N. L. Segal, "Monozygotic Triplets: Concordance and Discordance for Cleft Lip and Palate," *Twin Research and Human Genetics* 12 (2009): 403–406.

CHAPTER 2: THE SEED OF ALL THINGS

1. G. Y. Liu and D. Sabatini, "mTOR at the Nexus of Nutrition, Growth, Ageing and Disease," *Nature Reviews Molecular Cell Biology* 21 (2020): 183–203.

2. Z. Li et al., "Generation of Bimaternal and Bipaternal Mice from Hypomethylated ESCs with Imprinting Regions Deleted." *Cell Stem Cell* 23 (2018): 665–676.

3. L. Sagan, "On the Origin of Mitosing Cells," *Journal of Theoretical Biology* 14 (1967): 255–274.

4. D. A. Baum and B. Baum, "An Inside-Out Origin for the Eukaryotic Cell," *BMC Biology* 12 (2014): 76.

CHAPTER 3: A SOCIETY OF CELLS

1. A. Sebé-Pedros et al., "Early Evolution of the T-Box Transcription Factor Family," *Proceedings of the National Academy of Sciences of the United States of America* 110 (2013): 16050–16055.

2. D. Duboule, "The Rise and Fall of Hox Gene Clusters," *Development* 134 (2007): 2549–2560.

3. G. S. Richards and B. M. Degnan, "The Dawn of Developmental Signaling in the Metazoa," *Cold Spring Harbor Symposia on Quantitative Biology* 74 (2009): 81–90.

CHAPTER 4: REBIRTHS AND RESURRECTIONS

1. C. B. Fehilly, S. M. Willadsen, and E. M. Tucker, "Interspecific Chimaerism Between Sheep and Goat," *Nature* 307 (1984): 634–636.

2. J. B. Gurdon, T. R. Elsdale, and M. Fishberg, "Sexually Mature Individuals of *Xenopus laevis* from the Transplantation of Single Somatic Nuclei," *Nature* 182 (1958): 64–65.

3. I. Wilmut et al., "Viable Offspring Derived from Fetal and Adult Mammalian Cells," *Nature* 385 (1997): 810–813.

CHAPTER 5: MOVING PATTERNS

1. M. P. Harris et al., "The Development of Archosaurian First-Generation Teeth in a Chicken Mutant," *Current Biology* 16 (2006): 371–377; T. A. Mitsiadis, J. Caton, and M. Cobourne, "Waking Up the

Sleeping Beauty: Recovery of the Ancestral Bird Odontogenic Program," *Journal of Experimental Zoology* 306B (2006): 227–233.

2. M. G. Davey et al., "The Chicken talpid3 Gene Encodes a Novel Protein Essential for Hedgehog Signaling," *Genes & Development* 15 (2006): 1365–1377; K. E. Lewis et al., "Expression of ptc and gli Genes in talpid3 Suggests a Bifurcation in Shh Pathway," *Development* 126, no. 11 (June 1999): 2397–2407. doi: 10.1242/dev.126.11.2397.

3. L. Wolpert, "Positional Information and the Spatial Patterning of Cellular Differentiation," *Journal of Theoretical Biology* 25 (1969): 1–47.

CHAPTER 6: HIDDEN FROM VIEW

1. J. G. Dumortier et al., "Hydraulic Fracturing and Active Coarsening Position the Lumen of the Mouse Blastocyst," *Science* 365 (2019): 465–468.

2. S. F. Gilbert and R. Howes-Mischel, "'Show Me Your Original Face Before You Were Born': The Convergence of Public Fetuses and Sacred DNA," *History and Philosophy of the Life Sciences* 26 (2004): 377–394.

3. E. Sedov et al., "Fetomaternal Microchimerism in Tissue Repair and Tumor Development," *Developmental Cell* 20 (2022): 1442–1452.

4. M. Johnson, "A Short History of In Vitro Fertilization," *International Journal of Developmental Biology* 63 (2019): 83–92.

5. M. H. Johnson et al., "Why the Medical Research Council Refused Robert Edwards and Patrick Steptoe Support for Research on Human Conception in 1971," *Human Reproduction* 25 (2010): 2157–2174.

6. M. Roode et al., "Human Hypoblast Formation Is Not Dependent on FGF Signalling," *Developmental Biology* 361 (2012): 358–363; K. Niakan and K. Eggan, "Analysis of Human Embryos from Zygote to Blastocyst Reveals Distinct Expression Patterns Relative to the Mouse," *Developmental Biology* 375 (2013): 54–64.

7. A. McLaren, "Where to Draw the Line?" *Proceedings of the Royal Institution of Great Britain* 56 (1984): 101–121.

8. Ibid.

9. N. Hopwood, "Producing Development: The Anatomy of Human Embryos and the Norms of Wilhelm His," *Bulletin of the History of Medicine* 74 (2000): 29–79.

10. A. Poduri et al., "Somatic Activation of AKT3 Causes Hemispheric Developmental Brain Malformations," *Neuron* 74 (2012): 41–48; M. Lodato et al., "Aging and Neurodegeneration Are Associated with

Increased Mutations in Single Human Neurons," *Science* 359 (2018): 555–559.

11. Ibid.

12. S. Bizzotto et al., "Landmarks of Human Embryonic Development Inscribed in Somatic Mutations," *Science* 371 (2021): 1249–1253; S. Chapman et al., "Lineage Tracing of Human Development Through Somatic Mutations," *Nature* 595 (2021): 85–90; T. H. H. Coorens et al., "Extensive Phylogenies of Human Development Inferred from Somatic Mutations," *Nature* 597 (2021): 387–392.

CHAPTER 7: RENEWAL

1. L. Hayflick and P. S. Moorhead, "The Serial Cultivation of Human Diploid Cell Strains," *Experimental Cell Research* 25 (1961): 585–621.

2. A. Pozhitkov et al., "Tracing the Dynamics of Gene Transcripts After Organismal Death," *Open Biology* 7 (2017): 160267.

3. P. S. Eriksson et al., "Neurogenesis in the Adult Human Hippocampus," *Nature Medicine* 4 (1998): 1313–1317.

4. K. L. Spalding et al., "Retrospective Birth Dating of Cells in Humans," *Cell* 122 (2005): 133–143.

5. J. Till and E. A. McCulloch, "A Direct Measurement of the Radiation Sensitivity of Normal Mouse Bone Marrow Cells," *Radiation Research* 14 (1961): 1419–1430; A. Becker, E. McCulloch, and J. Till, "Cytological Demonstration of the Clonal Nature of Spleen Colonies Derived from Transplanted Mouse Marrow Cells," *Nature* 197 (1963): 452–454.

6. T. Sato et al., "Single Lgr5 Stem Cells Build Crypt-Villus Structures in Vitro Without a Mesenchymal Niche," *Nature* 459 (2009): 262–265.

7. G. Vlachogiannis et al., "Patient-Derived Organoids Model Treatment Response of Metastatic Gastrointestinal Cancers," *Science* 359 (2018): 920–926.

8. S. Yui et al., "Functional Engraftment of Colon Epithelium Expanded In Vitro from a Single Adult Lgr5[+] Stem Cell," *Nature Medicine* 18 (2012): 618–623.

9. K. Takahashi and S. Yamanaka, "Induction of Pluripotent Stem Cells from Mouse Embryonic and Adult Fibroblast Cultures by Defined Factors," *Cell* 25 (2006): 663–676; K. Takahashi et al., "Induction of Pluripotent Stem Cells from Adult Human Fibroblasts by Defined Factors," *Cell* 131 (2007): 861–872.

10. Ibid.

11. M. Eiraku et al., "Self-Organizing Optic-Cup Morphogenesis in Three Dimensional Culture," *Nature* 472 (2011): 51–56; T. Nakano et al., "Self-Formation of Optic Cups and Storable Stratified Neural Retina from Human ESCs," *Cell Stem Cell* 10 (2012): 771–785.

12. Ibid.

13. M. Lancaster et al., "Cerebral Organoids Model Human Brain Development and Microcephaly," *Nature* 501 (2013): 373–379.

14. X. Qian et al., "Brain-Region-Specific Organoids Using Mini Bioreactors for Modeling ZIKV Exposure," *Cell* 165 (2016): 1238–1254.

CHAPTER 8: THE EMBRYO REDUX

1. M. Bengochea et al., "Numerical Discrimination in *Drosophila melanogaster*," *BioRxiv* (2022) doi: https://doi.org/10.1101/2022.02.26.482107.

2. Y. Marikawa et al., "Aggregated P19 Mouse Embryo Carcinoma Cells as a Simple In Vitro Model to Study the Molecular Regulations of Mesoderm Formation and Axial Elongation Morphogenesis," *Genesis* 47 (2009): 93–106.

3. S. C. van den Brink et al., "Symmetry Breaking, Germ Layer Specification and Axial Organization in Aggregates of Mouse Embryonic Stem Cells," *Development* 141 (2014): 4231–4242.

4. L. Beccari et al., "Multi-axial Self-Organization Properties of Mouse Embryonic Stem Cells into Organoids," *Nature* 562 (2018): 272–276.

5. S. C. van den Brink, "Single Cell and Spatial Transcriptomics Reveal Somitogenesis in Gastruloids," *Nature* 582 (2020): 405–409; D. Turner et al., "Anteroposterior Polarity and Elongation in the Absence of Extraembryonic Tissues and of Spatially Localized Signalling in Gastruloids: Mammalian Embryonic Organoids," *Development* 144 (2017): 3894–3906.

6. B. Sozen et al., "Selg Assembly of Mouse Polarized Embryo-Like Structures from Embryonic and Trophoblast Stem Cells," *Nature Cell Biology* 20 (2018): 979–989; G. Amadei et al., "Inducible Stem-Cell Derived Embryos Capture Mouse Morphogenetic Events In Vitro," *Developmental Cell* 56 (2021): 366–382.

7. S. Tarazi et al., "Postgastrulation Synthetic Embryos Generated Ex Utero from Mouse Naïve ESCs," *Cell* 185 (2022): 3290–3306; G. Amadei et al., "Synthetic Embryos Complete Gastrulation to Neurulation and Organogenesis," *Nature* (2022). doi: 10.1038/s41586-022-05246-3.

8. N. Moris et al., "An In Vitro Model of Early Anteroposterior Organization During Human Development," *Nature* 582 (2020): 410–415.

9. T. Fulton et al., "Axis Specification on Zebrafish Is Robust to Cell Mixing and Reveals a Regulation of Pattern Formation by Morphogenesis," *Current Biology* 30 (2020): 2984–2994; A. Schauer et al., "Zebrafish Embryonic Explants Undergo Genetically Encoded Self Assembly," *Elife* 9 (2020): e55190. doi: 10.7554/eLife.55190.

10. L. Pereiro et al., "Gastrulation in Annual Killifish: Molecular and Cellular Events During Germ Layer Formation in *Austrolebias*," *Developmental Dynamics* 246 (2017): 812–826.

11. B. Steventon, L. Busby, and A. Martinez Arias, "Establishment of the Vertebrate Body Plan: Rethinking Gastrulation Through Stem Cell Models of Early Embryogenesis," *Developmental Cell* 56 (2021): 2405–2418.

12. J. B. A. Green and J. Sharpe, "Positional Information and Reaction Diffusion: Two Big Ideas in Developmental Biology Combine," *Development* 142 (2015): 1203–1211.

13. S. Niemann et al., "Homozygous WNT3 Mutation Causes Tetra-amelia in a Large Consanguineous Family," *American Journal of Human Genetics* 74 (2004): 558–563.

14. A. Deglincerti et al., "Self-Organization of the In Vitro Attached Human Embryo," *Nature* 533 (2016): 251–254.

15. A. Clark et al., "Human Embryo Research, Stem Cell–Derived Embryo Models and In Vitro Gametogenesis: Considerations Leading to the Revised ISSCR Guidelines," *Stem Cell Reports* 8 (2021): 1416–1424; R. Lovell-Badge, "Stem-Cell Guidelines: Why It Was Time for an Update," *Nature* 593 (2021): 479.

CHAPTER 9: ON THE NATURE OF A HUMAN

1. E. Haimes et al., "'So, What Is an Embryo?': A Comparative Study of the Views of Those Asked to Donate Embryos for hESC Research in the UK and Switzerland," *New Genetics and Society* 27 (2008): 113–126; I. de Miguel Beriain, "What Is a Human Embryo? A New Piece in the Bioethics Puzzle," *Croatian Medical Journal* 55 (2014): 669–671.

2. L. Yu et al., "Blastocyst-Like Structures Generated from Human Pluripotent Stem Cells," *Nature* 591 (2021): 620–626; X. Liu et al., "Modelling Human Blastocysts by Reprogramming Fibroblasts into iBlastoids," *Nature* 591 (2021): 627–632.

3. C. Zhao et al., "Reprogrammed Blastoids Contain Amnion-Like Cells but Not Trophectoderm," *BioRxiv* (2021). doi.org/10.1101/2021.05 .07.442980.

4. H. Kagawa et al., "Human Blastoids Model Blastocyst Development and Implantation," *Nature* 601 (2021): 600–605; A. Yanagida et al., "Naive Stem Cell Blastocyst Model Captures Human Embryo Lineage Segregation," *Cell Stem Cell* 28, no. 6 (2021): 1016–1022.

5. S. Tarazi et al., "Postgastrulation Synthetic Embryos Generated Ex Utero from Mouse Naïve ESCs," *Cell* 185 (2022): 3290–3306; G. Amadei et al., "Synthetic Embryos Complete Gastrulation to Neurulation and Organogenesis," *Nature* (2022). doi: 10.1038/s41586-022-05246-3.

6. A. K. Eicher et al., "Functional Human Gastrointestinal Organoids Can Be Engineered from Three Primary Germ Layers Derived Separately from Pluripotent Stem Cells," *Cell Stem Cell* 29 (2022): 36–51.

7. M. Saitou and K. Hayashi, "Mammalian In Vitro Gametogenesis," *Science* 374 (2021): eaaz6830.

8. J. Mench, "The Development of Aggressive Behaviour in Male Broiler Chicks: A Comparison with Laying-Type Males and the Effects of Feed Restriction," *Applied Animal Behaviour Science* 21 (1988): 233–242; Z. Li et al., "Genome Wide Association Study of Aggressive Behaviour in Chicken," *Scientific Reports* 6 (2016). doi:10.1038/srep30981. See also L. Trut, I. Oskina, and A. Kharlamova, "Animal Evolution During Domestication: The Domesticated Fox as a Model," *BioEssays* 31 (2009): 349–360.

9. P. Turley et al., "Problems with Using Polygenic Scores to Select Embryos," *New England Journal of Medicine* 385 (2021): 78–86.

10. D. Hong et al., "Initiating and Cancer-Propagating Cells in *TEL-AML1*-Associated Childhood Leukemia," *Science* 319 (2008): 336–339.

11. H. Driesch, *The Science and Philosophy of the Organism*, Gifford Lectures, 1907, vol. 1 (London: Adams and Charles Black, 1911).

12. M. Kirschner, J. Gerhart, and T. Mitchison, "Molecular Vitalism," *Cell* 100 (2000): 79–86.

EPILOGUE

1. C. DeLisi, "The Human Genome Project," *American Scientist* 76 (1988): 488–493.

2. R. Plomin, *Blueprint: How DNA Makes Us What We Are* (Cambridge, MA: MIT Press, 2018).

FURTHER READING

CHAPTER 1: NOT IN THE GENES

This chapter presents a brief account of how the gene has become so central to our lives and our culture. At the same time, it highlights the shortcomings of this point of view, especially when it comes to explaining how we are made. Mukherjee (2016) has produced an engaging narrative of the history of the gene that can be complemented by Zimmer's (2018) very different but thoughtful account of the consequences that the study of the gene has had for our history and daily lives. Mawer (2006) is a light exposition of the early days of genetics.

I have also included here some source references for the history of the genetics of various organisms that reveal how the many genes and acronyms that pervade the literature came to be. While most of the accounts focus on the magical powers of the gene and its effects on our lives, much less is known about the way this has come to be linked to embryos; details of how this happened can be found in flies (Gehring 1998, Gaunt 2019, Lipshitz 2005), fish (Nüsslein-Volhard 2012, Mullins et al. 2021), and mice (García-García 2020).

Significant contributions of women to the history of genetics are related in Korzh and Grunwald (2001), Richmond (2001), and Steensma, Kyle, and Shampo (2010).

Finally, it is not easy to find criticisms of the very gene-centered point of view, but D. Noble (2008) and Bahlla's (2021) comment on an account of the CRISPR story are good examples of attempts to do this. The book by Nurse (2020) is a summary of the current view of life from the perspective of genes and cells.

Bhalla, J. 2021. "We Haven't Really Cracked the Code." *Issues in Science and Technology* 37, no. 4. https://issues.org/code-breaker-doudna-isaacson-bhalla-review.

García-García, M. J. 2020. "A History of Mouse Genetics: From Fancy Mice to Mutations in Every Gene." In *Animal Models of Human Birth Defects*, edited by A. Liu. Advances in Experimental Medicine and Biology 1236. Singapore: Springer.

Gaunt, S. 2019. *Made in the Image of a Fly*. N.p.: Independently published.

Gehring, W. J. 1998. *Master Control Genes in Development and Evolution: The Homeobox Story*. New Haven, CT: Yale University Press.

Korzh, V., and D. Grunwald. 2001. "Nadine Dobrovolskaïa-Zavadskaïa and the Dawn of Developmental Genetics." *BioEssays* 23: 365–371.

Lipshitz, H. 2005. "From Fruit Flies to Fallout: Ed Lewis and His Science." *Developmental Dynamics* 232: 529–546.

Mawer, S. 2006. *Gregor Mendel: Planting the Seeds of Genetics*. New York: Harry N. Abrams.

Mukherjee, S. 2016. *The Gene: An Intimate History*. New York: Scribner.

Mullins, M., J. Navajas Acedo, R. Priya, L. Solnica-Kreel, and S. Wilson. 2021. "The Zebrafish Issue: 25 Years On." *Development* 148: 1–6.

Noble, D. 2008. *The Music of Life: Biology Beyond Genes*. Oxford: Oxford University Press.

Nurse, P. 2020. *What Is Life? Understand Biology in Five Steps*. Oxford: David Fickling Books.

Nüsslein-Volhard, C. 2012. "The Zebrafish Issue of Development." *Development* 139: 4099–4103.

Richmond, M. L. 2001. "Women in the Early History of Genetics: William Bateson and the Newnham College Mendelians, 1900–1910." *Isis* 92, no. 1: 55–90.

Steensma, D. P., R. A. Kyle, and M. Shampo. 2010. "Abbie Lathrop, the 'Mouse Woman of Granby': Rodent Fancier and Accidental Genetics Pioneer." *Mayo Clinic Proceedings* 85, no. 11: e83.

Zimmer, C. 2018. *She Has Her Mother's Laugh: The Powers, Perversions, and Potential of Heredity*. New York: Dutton.

CHAPTER 2: THE SEED OF ALL THINGS

Writing about something as visual as a cell, especially when it comes to describing what groups of cells can and do achieve, is a difficult task. Over the last few years, advances in microscopy have captured the life of a cell, often in real time, and revealed a world that we could not have imagined even twenty years ago. This makes this chapter challenging. I have tried to compensate for what can be a dry list of parts with images built up with words, but beyond a certain point, words simply cannot do justice to what we are seeing. The reader is encouraged to look at the many videos on the web: "The Inner Life of the Cell Animation" (www.youtube.com/watch?v=wJyUtbnOO5Y) and "Organelles of a Human Cell (2014) by Drew Berry and Etsuko Uno wehi.tv" (www.youtube.com/watch?v=2YCgro6BV8U) are beautiful visions of what is going on inside the confines of the plasma membrane. Alberts (2019) is a useful basic textbook for those interested in a slightly more detailed account of the science I introduce here.

The history of how we became aware of the existence and role of cells is discussed, in a somewhat scholarly manner, by Harris (1999), while the trailblazing work of Lynn Margulis and its implications are discussed in Sagan (2012), Gray (2017), and Sagan and Margulis (1992). Some of the ideas spun by her theory can be found in Baum and Baum (2020), Martijn and Ettema (2013), and Martin, Garg, and Zimorski (2015). The fascinating idea of the evolution of cell types is explored in Arendt (2008), and that of the enigmatic centriole, in Carvalho-Santos et al. (2011).

I also would like to highlight Bray's 2011 *Wetware*, a short, sharp account of the cell as a thinking entity, a view that, I suspect, will grow over the years. Biochemist Nick Lane (2022) explores aspects of metabolism now and in the origin of life in *Transformer*.

The central issue of *emergence* is difficult to grasp, and physicist Phil Anderson's 1972 essay on the subject is a standard reference for the curious mind.

Alberts, Bruce, Alexander Johnson, David Morgan, Karen Hopkin, Keith Roberts, Martin Raff, and Peter Walter. 2019. *Essential Cell Biology*. 5th ed. New York: W. W. Norton & Company.

Anderson, P. W. 1972. "More Is Different: Broken Symmetry and the Nature of the Hierarchical Structure of Science." *Science* 177: 393–396.

Arendt, D. 2008. "The Evolution of Cell Types in Animals: Emerging Principles from Molecular Studies." *Nature Reviews Genetics* 9: 868–882.

Baum, B., and D. A. Baum. 2020. "The Merger That Made Us." *BMC Biology* 18, no. 1: 72. doi: 10.1186/s12915-020-00806-3.

Bray, D. 2011. *Wetware: A Computer in Every Living Cell*. New Haven, CT: Yale University Press.

Carvalho-Santos, Z., J. Azimzadeh, J. B. Pereira-Leal, and M. J. Bettencourt-Dias. 2011. "Evolution: Tracing the Origins of Centrioles, Cilia, and Flagella." *Journal of Cell Biology* 194, no. 2: 165–175. doi: 10.1083/jcb.201011152.

Gray, M. W. 2017. "Lynn Margulis and the Endosymbiont Hypothesis: 50 Years Later." *Molecular Biology of the Cell* 28, no. 10: 1285–1287. doi: 10.1091/mbc.E16-07-0509.

Harris, H. 1999. *The Birth of the Cell*. New Haven, CT: Yale University Press.

Lane, N. 2022. *Transformer: The Deep Chemistry of Life and Death*. London: Profile Books.

Martijn, J., and T. J. G. Ettema. 2013. "From Archaeon to Eukaryote: The Evolutionary Dark Ages of the Eukaryotic Cell." *Biochemical Society Transactions* 41, no. 1: 451–457. doi: 10.1042/BST20120292.

Martin, W. F., S. Garg, and V. Zimorski. 2015. "Endosymbiotic Theories for Eukaryote Origin." *Philosophical Transactions of the Royal Society B: Biological Sciences* 370, no. 1678: 20140330. doi: 10.1098/rstb.2014.0330.

Sagan, D. 2012. *Lynn Margulis: The Life and Legacy of a Scientific Rebel*. White River Junction, VT: Chelsea Green.

Sagan, D., and L. Margulis. 1992. *Acquiring Genomes: A Theory of the Origin of Species*. New York: Basic Books.

CHAPTER 3: A SOCIETY OF CELLS

The transition from single cell to multicellular organisms was a major step in evolution and one around which there remain many questions and no clear explanation. The transition is associated with the emergence of pathways of cell communication and families of transcription factors linked to cell fates. From the cellular point of view the transition is associated with clonal multicellularity—that is, the emergence of diverse cell fates from one cell. For a view of the ongoing discussion of this process, see Ros-Rocher et al. (2021), Grosberg and Strathmann (2007), and Brunet and King (2017). Once multicellularity took hold of our imagination, scientists engaged in a phase of exploration described with great panache by S. J. Gould in 1990 in his classic *Wonderful Life*.

In this chapter I suggest that the emergence of multicellularity mounts a challenge not only to the gene-centric view of biology but, significantly, against what has been called "a gene's-eye view of evolution" put forward by Dawkins (1976, 1999) in his classic books and discussed by Agren (2021). Multilevel selection is explained in Szathmáry and Maynard Smith (1995). Martindale (2005) offers a very good discussion of the emergence of complex animals in the context of genes. Those interested in challenges to current views of evolution from the perspective of the mechanisms underlying the biology of cells should seek out Kirschner and Gearhart (2006).

Agren, A. 2021. *The Gene's-Eye View of Evolution*. Oxford: Oxford University Press.

Brunet, T., and N. King. 2017. "The Origin of Animal Multicellularity and Cell Differentiation." *Developmental Cell* 43: 124–140.

Dawkins, R. 1976. *The Selfish Gene*. Oxford: Oxford University Press.

Dawkins, R. 1999. *The Extended Phenotype: The Long Reach of the Gene*. Oxford: Oxford University Press.

Gould, S. J. 1990. *Wonderful Life*. New York: W. W. Norton & Company.

Grosberg, R. K., and R. R. Strathmann. 2007. "The Evolution of Multicellularity: A Minor Major Transition?" *Annual Review of Ecology, Evolution, and Systematics* 38: 621–654.

Kirschner, M., and J. Gearhart. 2006. *The Plausibility of Life: Resolving Darwin's Dilemma*. New Haven, CT: Yale University Press.

Martindale, M. Q. 2005. "The Evolution of Metazoan Axial Properties." *Nature Reviews Genetics* 6: 917–927.

Ros-Rocher, N., A. Perez-Posada, M. M. Leger, and I. Ruiz-Trillo. 2021. "The Origin of Animals: An Ancestral Reconstruction of the Unicellular-to-Multicellular Transition. *Open Biology* 11, no. 2: 200359. doi: 10.1098/rsob.200359.

Szathmáry, E., and J. Maynard Smith. 1995. "The Major Evolutionary Transitions." *Nature* 374: 227–232. doi: 10.1038/374227a0.

CHAPTER 4: REBIRTHS AND RESURRECTIONS

The mystery of what genes and cells have to do with each other at last began to be untangled with the cloning experiments of John Gurdon (2009) with frogs. The birth of Dolly the sheep extended Gurdon's findings to mammals and transformed what is essentially a scientific finding into popular folklore. Myelnikov and Garcia Sancho Sanchez (2017) offer an excellent account of this event, while Kolata (1997) provides a broader overview of cloning. These reprogramming experiments led to the uncovering of notions of genetic programs of development, as described by Peluffo (2015). Much of this discovery was developed in parallel with the rise of molecular biology, told in an engrossing manner by Judson in his classic 1996 book. Important parts of this story can be found in Jacob's (1988) autobiography, in which he gives a firsthand account of his groundbreaking interactions with Monod. The combination of cloning and molecular biology has created some interesting ideas, such as the notion of de-extinction, discussed in Shapiro (2015). The story of choosing the cells that gave rise to Dolly comes from Callaway (2016).

Waddington's implications for development are discussed in Moris, Pina, and Martinez Arias (2016), and its misunderstandings are brilliantly dissected in Pisco, Fouquier d'Herouel, and Huang (2016).

Callaway, E. 2016. "Dolly at 20: The Inside Story on the World's Most Famous Sheep." *Nature*. June 30. www.scientificamerican.com/article /dolly-at-20-the-inside-story-on-the-world-s-most-famous-sheep.

Gurdon, J. 2009. "Nuclear Reprogramming in Eggs." *Nature Medicine* 15: 1141–1144.

Jacob, F. 1988. *The Statue Within: An Autobiography*, translated by Franklin Philip. New York: Basic Books.

Judson, H. F. 1996. *The Eighth Day of Creation: The Makers of the Revolution in Biology*. Cold Spring Harbor, NY: Cold Spring Harbor Laboratory Press.

Kolata, G. 1997. *Clone: The Road to Dolly and the Path Ahead*. New York: William Morrow and Company.

Maienschein, J. 2014. *Embryos Under the Microscope*. Cambridge, MA: Harvard University Press.

Moris, N., C. Pina, and A. Martinez Arias. 2016. "Transition States and Cell Fate Decisions in Epigenetic Landscapes." *Nature Reviews Genetics* 17: 693–703.

Myelnikov, D., and M. Garcia Sancho Sanchez, eds. 2017. *Dolly at Roslin: A Collective Memory Event.* Edinburgh: University of Edinburgh.

Peluffo, A. E. 2015. "The Genetic Program: Behind the Genesis of an Influential Metaphor." *Genetics* 200: 685–696.

Pisco, A. O., A. Fouquier d'Herouel, and S. Huang. 2016. "Conceptual Confusion: The Case of Epigenetics." *BioRxiv.* doi: https://doi.org /10.1101/053009.

Shapiro, B. 2015. *How to Clone a Mammoth: The Science of De-extinction.* Princeton, NJ: Princeton University Press.

CHAPTER 5: MOVING PATTERNS

Discussions of the myth that storks deliver babies are all over the web; just ask Google, and you will be rewarded. The same is true of the variation on the story involving Paris. However, in the end, it all boils down to a story about eggs, the cell's ultimate masterpiece. M. Cobb (2007) tells the story of the search for the elusive mammalian egg. For an account of what embryos had to tell us before we became enamored of DNA instead, consult the classic *Ontogeny and Phylogeny* by S. J. Gould (1977) and Abzhanov (2013). Nonetheless, genes leave their mark on the mysteries of the development and evolution of embryos, as discussed in Duboule (2022), which also discusses the hourglass model (see also Richardson 1995).

At the center of how embryos build organisms is a process, gastrulation, and an idea, positional information. Both have been popularized by the great scientist and communicator Lewis Wolpert (1996, 2008). The story of his famous reference to gastrulation being more important than life and death is told in Hopwood (2022).

The story of the segmentation clock is considered in Pourquié (2022), and the very surprising appearance of teeth in birds, in Mitsiadis, Caton, and Cobourne (2006).

The science of embryos has its characters, though none is more charismatic and controversial than Ernst Haeckel. The works of Hopwood (2015), Richards (2008), and Richardson and Keuck (2002) trace the highs and the lows of this influential biologist. There are also good accounts of the famous Cuvier-Geoffroy debate (Appel 1987), the Haller–Wolff feud (Roe 1981), and the development of Spemann's school (Hamburger 1988).

Abzhanov, A. 2013. "Von Baer's Law for the Ages: Lost and Found Principles of Developmental Evolution." *Trends in Genetics* 29: 712–722.

Appel, T. 1987. *The Cuvier-Geoffroy Debate: French Biology in the Decades Before Darwin.* Oxford: Oxford University Press.

Cobb, M. 2007. *The Egg and Spoon Race.* New York: Simon & Schuster.

Duboule, D. 2022. "The (Unusual) Heuristic Value of Hox Gene Clusters: A Matter of Time?" *Developmental Biology* 484: 75–87.

Gould, S. J. 1977. *Ontogeny and Phylogeny.* Cambridge, MA: Harvard University Press.

Hamburger, V. 1988. *The Heritage of Experimental Embryology: Hans Spemann and the Organizer.* Oxford: Oxford University Press.

Hopwood, N. 2015. *Haeckel's Embryos.* Chicago: University of Chicago Press.

Hopwood, N. 2022. "'Not Birth, Marriage or Death, but Gastrulation': The Life of a Quotation in Biology." *British Journal for the History of Science* 55: 1–26.

Mitsiadis, T., J. Caton, and M. Cobourne. 2006. "Waking-Up the Sleeping Beauty: Recovery of the Ancestral Bird Odontogenic Program." *Journal of Experimental Zoology* 306B: 227–233.

Pourquié, O. 2022. "A Brief Story of the Segmentation Clock." *Developmental Biology* 485: 24–36.

Richards, R. J. 2008. *The Tragic Sense of Life: Ernst Haeckel and the Struggle over Evolutionary Thought.* Chicago: University of Chicago Press.

Richardson, M. 1995. "Heterochrony and the Phylotypic Period." *Developmental Biology* 172: 412–421.

Richardson, M., and G. Keuck. 2002. "Haeckel's ABC of Evolution and Development." *Biological Reviews* 77: 495–528.

Roe, S. A. 1981. *Matter, Life, and Generation: 18th Century Embryology and the Haller–Wolff Debate.* Cambridge: Cambridge University Press.

Wolpert, L. 1996. "One Hundred Years of Positional Information." *Trends in Genetics* 12: 359–364.

Wolpert, L. 2008. *The Triumph of the Embryo.* Oxford: Oxford University Press.

CHAPTER 6: HIDDEN FROM VIEW

Difficult as human development is to observe inside the womb, the last 150 years have seen significant progress in documenting this story. The work of Lynn Morgan (2009) is essential reading on this topic as it recounts much of this and places it within a social and anthropological context. Nilsson (1965) put images of embryos and fetuses in the public imagination.

The reports on the work of the Warnock Committee are worth a look for interested readers (Warnock 1984, 1985).

Morgan, L. 2009. *Icons of Life: A Cultural History of Human Embryos.* Berkeley: University of California Press.

Nilsson, L. 1965. *A Child Is Born.* New York: Dell.

Warnock, M. 1984. *Report of the Committee of Inquiry into Human Fertilization and Embryology.* London: Her Majesty's Stationery Office.

Warnock, M. 1985. "The Warnock Report." *British Medical Journal* 291, no. 6493: 489.

CHAPTER 7: RENEWAL

The last few years have seen a revolution in biology. In the same manner that in the 1970s and 1980s the molecular analysis of living systems transformed our understanding of development, disease, and evolution, the start of the twenty-first century has recognized the cell as the driving force behind and protagonist within these processes. For additional detail about the cell, it would be wise to return to the references for Chapters 2 and 3.

The development of technologies associated with cells in culture is told in Landecker (2007), and Witkowski (1980) goes through the details of the intriguing story of A. Carrell and his immortal cells. Skloot (2010) tells the story of Henrietta Lacks and how her cells changed research in cell biology and made us aware of their relationship to us as individuals. Grimm (2008) tells the fascinating story of how nuclear weapons tests came to the assistance of biology.

The story of stem cells is discussed in Maehle (2011), and the path to the discovery of embryonic stem cells is well summarized by Martin Evans (2011). This will allow you, if you are interested, to dip into literature on the current progress in the use of these cells to build organs and tissues (Corsini and Knoblich 2022; Dutta, Heo, and Clevers 2017; Sasai, Eiraku, and Suga 2012).

An excellent, accessible account of the development of efforts to use cells to understand how tissues and organs are made can be found in Ball (2019).

Ball, P. 2019. *How to Grow a Human.* Chicago: University of Chicago Press.

Corsini, N. S., and J. Knoblich. 2022. "Human Organoids: New Strategies and Methods for Analyzing Human Development and Disease." *Cell* 185: 2756–2769.

Dutta, D., I. Heo, and H. Clevers. 2017. "Disease Modeling in Stem Cell– Derived 3D Organoid Systems." *Trends in Molecular Medicine* 23: 393–410.

Evans, M. 2011. "Discovering Pluripotency: 30 Years of Mouse Stem Cells." *Nature Reviews Molecular Cell Biology* 12: 680–686.

Grimm, D. 2008. "The Mushroom Cloud's Silver Lining." *Science* 321: 1434–1437.

Landecker, H. 2007. *Culturing Life: How Cells Became Technologies.* Cambridge, MA: Harvard University Press.

Maehle, A. H. 2011. "Ambiguous Cells: The Emergence of the Stem Cell Concept in the Nineteenth and Twentieth Centuries." *Notes and Records of the Royal Society of London* 65: 359–378.

Sasai, Y., M. Eiraku, and H. Suga. 2012. "In Vitro Organogenesis in Three Dimensions: Self-Organizing Stem Cells." *Development* 139: 4111–4121.

Skloot, R. 2010. *The Immortal Life of Henrietta Lacks.* New York: Crown.

Witkowski, J. A. 1980. "Dr. Carrell's Immortal Cells." *Medical History* 24: 129–142.

CHAPTER 8: THE EMBRYO REDUX

The ability of cells to recapitulate many steps of the development of embryos in vitro is a novel and promising area of research, but there have been few accounts of this field to date. Some of this story can be found in Ball's (2019) *How to Grow a Human*, already referred to earlier.

Zernicka-Goetz and Highfield (2020) provide a personal account of attempts to build embryos from stem cells. For more technical discussions, see Martinez Arias, Marikawa, and Moris (2022), Rivron et al. (2018), and Shahbazi, Siggia, and Zernicka-Goetz (2019).

The website of the Institut de la Vision (https://transparent-human -embryo.com) offers an immense and beautiful resource for the exploration of the human embryo.

Ball, P. 2012. *Unnatural: The Heretical Idea of Making People.* New York: Vintage Books.

Martinez Arias, A., Y. Marikawa, and N. Moris. 2022. "Gastruloids: Pluripotent Stem Cell Models of Mammalian Gastrulation and Body Plan Engineering." *Developmental Biology* 488: 35–46.

Rivron, N., J. Frias-Aldeguer, E. J. Vrij, J.-C. Boisset, J. Korving, J. Vivié, R. K. Truckenmüller, A. van Oudenaarden, C. A. van Blitterswijk, and N. Geijsen. 2018. "Blastocyst-like Structures Generated Solely from Stem Cells." *Nature* 557: 106–111.

Shahbazi, M., E. Siggia, and M. Zernicka-Goetz. 2019. "Self-Organization of Stem Cells into Embryos: A Window on Early Mammalian Development." *Science* 364: 948–951.

Zernicka-Goetz, M., and R. Highfield. 2020. *The Dance of Life*. New York: Penguin.

CHAPTER 9: ON THE NATURE OF A HUMAN

The developments that have led to the generation of embryo-like structures from embryonic stem cells have raised questions not only about the nature and workings of embryos but also about the very essence of human beings. Maienschein (2005) touches upon some of the more accessible dimensions of these discussions. The proceedings of the Nuffield Council on Bioethics (2017) contain a number of interesting contributions from all sides of the debate and make for valuable reading. On more material aspects of the problem, Jamie Davis's 2015 book offers an excellent account, while the collection by Hopwood, Flemming, and Kassell (2018) is a useful historical reference. Franklin (2013) is a thorough discussion of issues of embryos and relatedness.

Issues of human genetic diversity are clearly discussed in Rutherford (2016). While it is difficult to find an accessible nonspecialist account of the recent and complex matter of polygenic scores, Torkamani, Wineinger, and Topol (2018) might provide a helpful guide. The publication of Harden (2021) prompted extensive discussion, and I would encourage you to read it and make up your own mind about the meaning of these scores. Finally, the work of Sapp (2003) delivers a clear and direct exposition of the evolution of biological ideas across the last two centuries.

Davis, J. 2015. *Life Unfolding*. Oxford: Oxford University Press.

Franklin, S. 2013. *Biological Relatives: IVF, Stem Cells, and the Future of Kinship*. Durham, NC: Duke University Press.

Harden, K. P. 2021. *The Genetic Lottery: Why DNA Matters for Social Equality*. Princeton, NJ: Princeton University Press.

Hopwood, N., R. Flemming, and L. Kassell, eds. 2018. *Reproduction: Antiquity to the Present*. Cambridge: Cambridge University Press.

Maienschein, J. 2005. *Whose View of Life? Embryos, Cloning, and Stem Cells*. Cambridge, MA: Harvard University Press.

Nuffield Council on Bioethics. 2017. *Human Embryo Culture: Discussions Concerning the Statutory Time Limit for Maintaining Human Embryos in Culture in the Light of Some Recent Scientific Discoveries*. London: Nuffield Council on Bioethics.

Rutherford, A. 2016. *A Brief History of Everyone Who Ever Lived: The Stories in Our Genes*. London: Weidenfeld & Nicolson.

Sapp, J. 2003. *Genesis: The Evolution of Biology*. Oxford: Oxford University Press.

Torkamani, A., N. E. Wineinger, and E. Topol. 2018. "The Personal and Clinical Utility of Polygenic Risk Scores." *Nature Reviews Genetics* 19: 581–591.

EPILOGUE

It looks as if, finally, the cell is showing its head above the parapet of the genome, and 2022 saw the publication of Siddhartha Mukherjee's *The Song of the Cell*, which leans in the direction of the argument here, advocating a need to embrace and explore what cells have to offer in terms of the explanations for and causes of much of our biology.

Mukherjee, Siddhartha. 2022. *The Song of the Cell*. New York: Scribner.

INDEX

ALFONSO MARTINEZ ARIAS is ICREA Research Professor at Universitat Pompeu Fabra in Barcelona and was previously professor of developmental mechanics at the University of Cambridge. For his contributions to developmental biology, he has received the Waddington Medal of the British Society of Developmental Biology and the Lifetime Achievement Award of the Society of Spanish Researchers in the United Kingdom. As coauthor of *Principles of Development* along with Lewis Wolpert and Cheryll Tickle, he received the Royal Society of Biology Undergraduate Textbook Prize. He lives in Barcelona, Spain.

.